ANDREA BAUER is an innovation designer, tech philosopher and author. She is the founder of BEAM Studio, an innovation firm that applies novel technologies and methods, to set-up spaces and processes to create cutting-edge services, products, and business models. Always fascinated by the question how technology can improve our everyday lives, she follows a purpose-driven focus, to accelerate her efforts towards a more responsible future.

www.andrea-bauer.com

Andrea Bauer

Trusting Mobile Payment

How the Trust-Factor forms the Mobile Payment Process

Author: Andrea, Bauer
Scientific Committee:
Prof. Dr. Sabine Fischer, University of Arts, Berlin
Prof. Dr. Gesche Joost, Technical University, Berlin

Acknowledgments:
Dr.-Ing. Zhiyun Ren, Deutsche Telekom AG
Marc-André Fengler, Vodafone D2 GmbH
Holger Spielberg, Pay Pal Deutschland GmbH

© 2013 Andrea Bauer
Publisher: tredition GmbH, Hamburg
ISBN (Paperback): 978-3-7439-3423-8
ISBN (Hardcover): 978-3-7439-3406-1
ISBN (E-Books): 978-3-7439-3558-7

Printed in Germany

The work, including its parts, is protected by copyright. Any reproduction without permission of the publisher and the author is illegal. This is especially true for electronic or other reproduction, translation, dissemination, and public disclosure.

Bibliographic information published by the National Library:
The German National Library lists this publication in the Deutschen Nationalbibliografie. Detailed bibliographic data are available on the Internet at http://dnb.d-nb.de.

For my beloved family and adored friends.

Index

1 Introduction 11

2 Theoretical Consideration 16
2.1 About Trust 17
2.1.1 Money and Trust 17
2.1.2 Trust: An Inter-Disciplinary Consideration 21
2.1.3 Functions of Trust 25
2.1.3.1 Reduction of Complexity 26
2.1.3.2 Familiarity and Distrust 28
2.1.3.3 Interpersonal and Systemic Trust 31
2.1.4 Determinants of Trust 36
2.1.4.1 Classification 36
2.1.4.2 Habitus 38
2.1.4.3 Reputation 39
2.1.5 Creating Trust 41
2.1.5.1 Phases of Trust 41
2.1.5.2 The Economic Decision-Making Process 44
2.1.5.3 Technology Acceptance Model 48
2.1.5.4 Tech Trust as Qualified Reliability 51
2.1.6 Summary 54
2.2 About Mobile Payment 57
2.2.1 Terminology 60
2.2.1.1 Definition 60
2.2.1.2 Delineation 61
2.2.2 Evolution of the Ecosystem 61
2.2.2.1 Transmission Technology 62
2.2.2.2 Smartphones and Usability 65
2.2.2.3 Mobile App Boom 68
2.2.2.4 Market Acceptance 72
2.2.3 Classification 79
2.2.3.1 Place of Payment 79
2.2.3.2 Payable Amount 82
2.2.3.3 Time of Debit 83
2.2.3.4 Place of Application 86
2.2.3.5 Applications in Commerce 88

2.2.4 Mobile Payment Process	91
2.2.4.1 Agents	91
2.2.4.2 Value-Added Chain	93
2.2.4.3 Remote and Proximity	96
2.2.5 Stakeholders	98
2.2.5.1 Customers	99
2.2.5.2 Merchants	101
2.2.5.3 Financial Service Providers	102
2.2.5.4 Mobile Phone Providers	103
2.2.5.5 Technology Providers	105
2.2.5.6 Specialized Mobile Payment Start-ups	109
2.2.6 Summary	111
3 Practical Comparisons	**113**
3.1 Introduction	114
3.2 Telekom	116
3.2.1 Deutsche Telekom's Payment Ecosystem	116
3.2.2 The Mobile Wallet	118
3.2.3 An User-Oriented Approach	120
3.2.4 The Security Aspect	122
3.2.5 Trust Threshold	125
3.3 Vodafone	127
3.3.1 The Company	127
3.3.2 Evolution of Mobile Payment	128
3.3.3 Systemic and Perceived Security	130
3.3.4 On Establishing Trust	132
3.4 PayPal	136
3.4.1 The Company	136
3.4.2 Mobile Payment: A Center Piece	138
3.4.3 On Trusted Intermediary	141
3.4.4 Mobile First	143
3.4.5 Trust in Communication	145
4 Conclusion	**148**
Endnote	**156**

1 Introduction

The rapid technological development and global distribution of wireless devices opens up new possibilities for communication and business activity in our global society. Mobile terminals, such as cell phones, PDAs, smart phones and tablet PCs have become the mainstay of post-modern communication. Mobile users can employ digital services independent of location and time. Wireless devices allow the flexible utilization of digital services that were once limited to stationary use on a personal computer (PC). The range of potential applications has grown due to improved technology and processes of the mobile ecosystem. The wireless device has become a multi-functional machine. Besides the classical telephone service, a large variety of new services in the areas of navigation, social media, photography and alarms have been added to what was once a mobile telephone.

Adequate mobile payment services are now needed to monetize those services. The idea of using a cell phone as a payment medium is not new. The mobile payment topic has been at the forefront since the 1990s, especially due to mobile value added services (VAS).

However, only few of the mobile payment solutions reached a significant market maturity or relevance. Apart from many other success factors, the use of such services requires a great deal of trust by the user towards the technical set-up, procedural information and data processing. Especially complex payment services require particular attention on the trust building process. But to build trustworthiness it is not enough to only provide a secure data transaction and storage. Trust must also be supported through user experience design, transparent communication and imparted information so the user easily learns, understands, and eventually accepts the service. Enduring success for a mobile payment service can only be achieved by establishing a long-term relationship between the user and the system based on trust.

Trust is an essential factor for any user to select a product or service. Whether and how the trust factor was taken into account and represented when service provider create a mobile payment service has not yet been scientifically analysed. With this paper this gab will be closed. It will analyse the role of trust in the construction and communication of a mobile payment services from the perspective of the service providers. In order to answer this question acknowledged sociological theories on trust will be examined and conceptually expanded.

The paper aims to create a general understanding of the relevance of the trust-factor for mobile payment services by means of a practical comparison in order to elaborate, which aspects of trust are particularly relevant, and in what way they should be considered and designed. In conclusion, open research questions will be determined and an outlook will be delivered.

The purpose of this paper is not to develop a new trust model for mobile payment service, nor to combine or dismiss existing services. Its purpose rather is to analyse, understand, and reflect the inherent trust-related tension between people and mobile payment services on the basis of existing models. Specifically it will focus on trust towards systems, as the user always has to verify the trustworthiness of a technical system.

Consequently, this paper focuses purely on mobile payment and on the transfer of monetary value. This means it will not answer the question about trust in the monetary system, as it will not examine the trustworthiness of the value of the currency itself, but its transfer.

A theoretical structure is established in the first part of the paper in order to answer these questions. It is based on research of literature, and it provides an explanation of the fragmented concept of trust and the individual description of topic-related models and definitions. It specifically covers statements by authors who have created a reference to describe systems or money. The following chapter explains mobile payment and provides profound insight into its facets. In addition to a general delimitation of the concept, its development, applications, and characteristics will be discussed.

The second part of the book deals with the practical comparison. Therefor the consideration of aspects of trust in the construction and communication of existing mobile payment services are examined by conducting expert interviews. In this case one-on-one expert interviews are preferred to avoid group dynamics when it comes to enquire opinions. The experts are interviewed verbally. To obtain the most authentic information, the interview is to be conducted in the regular working environment. As this study is more of a hermeneutic exploration, a semi-standardized questionnaire is applied to concede space for additional information and/or questions. This grants the interviewees more freedom for their replies, and they can go deeper as they deem appropriate.

Interviews are conducted with representatives of three current mobile payment providers. In order to ensure a certain degree of comparability of the data, a standardized questionnaire or conversation guide has been developed. It contains main questions and supporting sub-questions. As it is a semi-open interview, the main questions are always asked, while the supporting questions serve to maintain an organic conversation. Besides the interview guide, an audiotape will be used to record the conservations.

2 Theoretical Considerations

2.1 About Trust

Trust is a complex social phenomenon often paraphrased with words such as confidence, familiarity, credibility, or trustworthiness.

This paper does not pretend to extensively depict all scientific approaches to trust. Instead, the phenomenon trust is to be examined in the context of mobile payment services, and its aspects and characteristics are to be analysed. The following text therefore concentrates on the description of the functions, determinants, and peculiarities of trust.

The goal is to find an answer to the following questions: What is the relationship between trust and payment processes? What is the interdisciplinary definition of trust? What is the nature of trust? What are its forms and scientific manifestations?

2.1.1 Money and Trust

This paper particularly focuses on the payment process. The payment process represents the monetary transaction process between two parties. In order to understand the relationship between trust and paying, we must first explain the fundamental relationship between trust and money.

The transaction process illustrates the central nature of money as a social contract and transmitter of economical interactions. Especially this social interaction needs to bridge this moment of economical uncertainty. Trust becomes equal to a social advance.

This way the use of money is historically based on designing the trust elements of a transaction process, provided by a current social monetary system. This means the instant of payment only acquires its right to exist by successfully creating trust between the dealing parties. Luhmann says: "Those who believe in the stability of monetary value and the continuity of a variety of possibilities of usage, essentially assume that a system functions, and they don't confide in people they know, but in this functioning."[1]

The bearer of the money symbol, i.e., the coin or bill, usually doesn't have the same intrinsic value as the imprinted face value suggests. Without trust, what is left is only the intrinsic substance value of the metal or paper, which frequently is insignificant compared to the face value.

A high relevance of trust is apparent upon the close consideration of the labelling and design of bills: "I promise to pay the bearer on demand the sum of...," reads the bill of the Bank of England. This indicates that the bill bearing the money symbol represents a physical promise for usage of the exchanged value through the paper bill elsewhere.

---

[1] Cf. Luhmann, 2000, p. 64

The US $10 bill also shows trust-related characteristics: "In God We Trust," it says on the reverse of the bill. But the person that we are actually supposed to trust is the U.S. Treasury Secretary. Alexander Hamilton, the first Treasury Secretary of the United States and co-founder of the modern banking system, is still featured on today's US $10 bill. On the other hand, when Germans trade their work or merchandise against euros, they trust the President of the European Central Bank (ECB), Mario Draghi (since November 2011). The ECB, together with the national central banks, is responsible for the control and supervision of the banking system in the euro area. On the face of any euro bill, we find the signature of the incumbent President of the ECB as a symbol of trustworthiness and validity of the paper.

Money obtains its role and necessity due to the existence of an intrapersonal idea of values, and it manifests in the interpersonal exchange of values. Money, as the physical bearer of the money symbol, thus becomes the central element of the payment process. During the physical moment of payment with cash, trust is primarily generated through the forgery-proof quality of the means of payment.

Besides the ECB President's imprinted signature, the euro bill contains a large number of security features, such as watermarks, security thread, recess printing, see-through register, micro-printing, infrared features, black light features, magnetic security detection, pearlescent strip, a special foil element with holograms, bar code, and colour shifting effects when the bill is tilted.[2]

Today cash is one of many means of payment. Cashless payments have spread swiftly in the past few years and steadily increased its density of use.[3] Compared to digital money, cash has expensive disadvantages, such as the cost of its production and provision, the cost of withdrawing it for the end consumer, or the cost of logistics for merchants in daily business.

In contrast, cashless payment processes often seem less expensive and more convenient. However, with the digitalization of the payment processes, the bearers of the money symbol change as well. They show themselves in their new form as a plastic card or as digital information on a cell phone. But even if their shape changes, the user's claim for trust in the bearer and the payment process of the money persist. Although the trust-promoting aspects are no longer as visible as on the bill, the service provider still needs to consider them as inherent elements of the process or medium.

---

[2] Cf. German Central Bank: Informationen über die Bundesbank Sicherheitsmerkmale der Euro-Banknoten.(Information on the Security Features of Euro Bills)
[3] Cf. Statista: Annual growth rates of cashless transactions 2001-2007 based on the 10 largest cashless markets.

2.1.2 Trust: An Inter-Disciplinary Consideration

First the term trust must be explained. This word is used so naturally, yet so hard to grasp. A general definition of trust is not available in scientific literature. There seems to be consensus that trust is an immeasurable value for an interpersonal relationship in society. There is also agreement that building trust is an active process and that relationships based on trust are desirable.

Etymologically considered, the origin of the German noun *Vertrauen* (trust) and the German verb *vertrauen* (to trust) is the Middle High German word *truwen*.[4] The verb *truwen* was used during the Middle Ages in the sense of believe, hope, and give credit. Thus, the German terms for believe and trust have a strong etymological relationship. The German verb *glauben* (believe) has its origin in the Gothic word *galaubjan*.[5] *Galaubjan* had, especially with the pagan Germans, the meaning of the friendly trust of a person towards the gods.

However, the usage of the term moved further and further away from its description of a godly relationship and found more application in secular, social relationships. Today, trust describes the

---

[4] Cf. Bentele, 1993, p. 305
[5] Cf. Bentele, 1993, p. 305

subjective conviction that something is right or true, according to Schweer & Thies.[6]

Since trust, as an informal, intangible mechanism, is an essential social element, it has been a popular research topic in various social areas and scientific disciplines. The result of this is a large variety of scientific descriptions, approaches, and classifications.

In psychology, the topic of trust can look back on a long and comprehensive research history in several sub-disciplines, such as personality psychology, psychoanalysis and depth psychology.[7] Here it is important to note that despite the different approaches, the focus has always been on the individual in order to investigate the explicit understanding of trust or the determinants of a trusting attitude.

Deutsch, for instance, describes trust as a recurring, conscious, voluntary, observable behavioural characteristic. In his view, trust is a moment during which the vulnerability of the trust giver increases as against the trust taker, as the trust giver cannot control the trust taker's future behaviour.[8]

---

[6] Cf. Schweer/Thies, 2003, p. 15
[7] Cf. Thomas, 2005, p. 2
[8] Cf. Deutsch, 1962, p. 302 ff.

In philosophy, the term trust has also been discussed for a long time. Social interactions are the focus of philosophical research. Trust is therefore discussed on a higher meta-level, and understanding the value and meaning of the mechanism is one of the major goals. This means trust is mainly conceived as an ethical discourse.[9]

Besides the moral-philosophical approach, there are also examples of explaining trust in existential and social philosophy.[10] Even if the term trust is described in a similar way as in other disciplines, the philosophical approach lacks a rational, measurable component. The highly abstract discourse around the topic will only have limited importance for this investigation.

Unlike psychology and philosophy, sociology puts the relationship between the individual and society in the center of investigation. Along with decision and structure-theoretical approaches, the system-theoretical approaches in particular take a relevant position for this book.[11]

Simmel, for example, describes trust as a combination of knowledge and non-knowledge.[12] Luhmann, the most important and most famous

---

[9] Cf. Evanoff, 2006, p. 421 ff.
[10] Cf. Brieskorn, 2009, p. 39
[11] As to general fundamentals and approaches of sociology, cf. Esser 2002 and Miebach 2006
[12] Cf. Simmel, 1922, p. 263 f.

sociologist and representative of the systems theory, considers trust as a mechanism to reduce social complexity.[13] Just as Giddens, he also distinguishes between trust in individuals and trust in abstract systems.[14]

Admittedly, sociology also still seems to depict the term trust very vaguely. Nevertheless, the sociological concept of the term is the most significant for the present analysis. Especially the systemic contemplation and the interpretation of the determinants and functions of trust by this discipline help analyse and better understand the mechanism of trust between user and mobile payment service. This is why the following paragraphs cover more details of the approaches on trust by Luhmann, Misztal, and Lewicki & Bunker.

Even if the trust-factor doesn't have a long tradition in the economics of technology, it is attracting ever more attention. This is also reflected by the theme of the CeBIT in 2012: "Managing Trust".[15] Within the framework of economics, the investigation of the trust mechanism is often focused on relationship management, both within and outside a company.[16] This aspect has been applied in the

---

[13] Cf. Luhmann, 2000, p. 27 ff.
[14] Cf. Giddens, 1995b, p. 107; Luhmann, 2000, p. 60 ff.
[15] Cf. CeBIT leading theme 2012: *Managing Trust*, Statements by the Industry on *Managing Trust*. Trust and security in the digital world.
[16] Cf. Bruhn/Esch/Langner, 2009, p. 487 ff.

context of transaction cost theory for a long time.[17] "Theoretically, it is argued that generalized trust functions as a cultural resource, which makes economic exchange and transactions more productive by allowing for more and more encompassing actions (networking) by reducing transaction costs and costly controls as well as enhancing the flow of information."[18]

Therefore, the research of trust in the network of relationships of users and mobile payment services turns to two scientific approaches in this paper. On one hand, it considers the technology acceptance model for mobile services, and on the other, it addresses Ed Gerck's theory. He defines the aspect of trust within the framework of Internet communication as qualified reliability.

2.1.3 Functions of Trust

Nikolas Luhmann is arguably one of the most important German sociologists of the 20th century. His works have been applied in several different scientific disciplines. Luhmann is known as one of the founders of the sociological systems theory. He analyses the main topics of life, such as trust, power, and love in a functionalistic manner. This paper focuses on the analysis of trust he formulated in the 1970s. For this purpose, Luhmann chose a heuristic,

---

[17] Cf. Loose/Sydow, 1994, p. 162 f.
[18] Volken, 2002, p. 1

functional method of analysis that concentrates on the derivation of solutions on the basis of problems.

Trust is an elementary matter of fact in social life and thus an ever-present phenomenon, according to Luhmann. Due to the virtual nature of trust, it appears in the widest variety of contexts, making it nearly impossible to analyse. In order to understand the phenomenon of trust despite this fact, Luhmann examines the underlying functions of trust to make it more palpable. The following chapters will deal with the main functions of trust: reduction of complexity, familiarity, distrust, as well as person- and system-related trust.

2.1.3.1 Reduction of Complexity

The expansion of information technologies, particularly of digital and mobile media, massively increased the volume of information and thus the complexity of the perceived environment.[19] The Bielefeld sociologist suitably called the mechanism for the reduction of social complexity the central function of trust.[20] The complexity of the social structure is growing especially in modern societies with extreme division of labour, a high degree of specialized knowledge and a steadily rising spread of digital, bi-directional media. This is why our own knowledge, resources, and capacities are no longer sufficient for making decisions based on acquired

---

[19] Cf. Fukuyama, 1995, p. 3 ff.
[20] Cf. Luhmann, 2000, p. 27 ff.

and reflected knowledge and participating in social life at the appropriate pace. Several different studies confirm a real explosion of global knowledge through digital media.[21] Recorded knowledge grew by 30% from 1999 to 2000.[22] At the same time, a long-term study in Germany shows that the time of interaction with media doesn't shift through the increase in digital options for use, but it rises overall. In the age group of 14 to 49, the average duration of media use even climbed to eight hours per day.[23]

In the case of existing complexity, this always carries a risk of poor performance or disappointment after a decision to use is made. During those moments, trust becomes a bet on the future, while at a positive decision moment there must be an excess of certainty in order to generate a positive result.

The term "risk society" was created at the beginning of the 1990s, when increasing social complexity was being observed. The term describes the replacement of modernism as an industrial society by postmodernism as a risk society. It describes the phenomenon of a positive correlation between

---

[21] On the history of knowledge cf. Stuhlhofer, 1983. On the explosion of knowledge through digitalization cf. Hilbert/López (01.04.11): The World's Technological Capacity to Store, Communicate, and Compute Information

[22] Cf. o.V. (2003): How Much Information?, p. 2

[23] Cf. SevenOne Media (03.12.03): Long-term study of media use in Germany.

wealth production and risk production.[24] The number of options with different risks grew with the transition into the chaotic, multi-directional postmodern era. The consequence was permanent decision pressure on the individual. The evolution of human cognition was progressing to a much lesser extent. Trust thus becomes a person's significant capability to efficiently select and rapidly rate promising options. By implication this means that wherever there is trust, there must also be a risk as well as social complexity, which entails a variety of possible experiences.[25]

The novelty, the large range of choices, and the complexity of mobile payment services make it almost impossible and often too cost and time-intensive to extensively verify the service before use. And so trust becomes a social advance service here as well; it overcomes an alleged complexity of the data to be checked with its inherent risk, and allows for faster ability to act.

2.1.3.2 Familiarity and Distrust

Petermann describes trust as the bridging of lack of knowledge by a trust giver as to the future behaviour of the object of trust, with simultaneous positive expectations of the future.[26]

---

[24] Cf. Beck, 1992, p. 19 ff.
[25] Cf. Luhmann, 2000, p. 8
[26] Cf. Petermann, 1996, p. 15

Trust is also a positive projection into the future based on our experiences in the past. "Who grants trust, anticipates the future," Luhmann says.[27] However, while trust is linked to the future, familiarity is related to the past. The things learned and reflected on in the past shape the identity that holds the prospect of possible future experiences. Hence the processing of experience becomes the basis of trust building.[28]

However, trust can only be triggered and built in the present. For that the necessary inner security is required. Inner security is generated by the generalization of expectations based on past experience.[29]

Luhmann regards the approximation of relationships as a continuous process consisting of learning and testing trust relationships.[30] The perceived data of the object of trust is constantly being processed and compared to the inner order and expectations.[31] The symbolic information of the object of trust serves as a sign of the familiar. Inner security is created when symbols serve positive expectations, while the perceived risk seems isolatable. Rotter also describes an approach of psychology of learning, which depicts trust as

---

[27] Luhmann, 2000, p. 9
[28] Cf. Luhmann, 2000, p. 13 ff.
[29] Cf. Mead, 1938, p. 175
[30] Cf. Luhmann, 2000, p. 29
[31] Cf. Luhmann, 2000, p. 32

subjective expectations based on the reliability of promises.[32]

There is constant control of the information between the individual and the object of trust per feedback loop in order to learn whether the trust is still justified.[33] Trust is of a very fragile nature. Trust could also be compared to a loan in the sense of an advance payment, within the framework of which unfavourable experiences can be interpreted favourably or be absorbed. Nonetheless, there are thresholds of trust-critical behaviour that lead to abrupt distrust or, to stay with the loan metaphor, exhaust the credit limit.[34]

This psychological threshold is subjectively variable, and when crossed, it leads to immediate reorientation.[35] The threshold moment is not reached suddenly. It is built up over time. It is preceded by an accumulation of many negative experiences which pass through phases of justification, indulgence, and caution until eventually a stage of distrust is reached.[36] The change from trust to distrust happens very quickly. Extremely time-consuming efforts are necessary to convert distrust back to trust.

---

[32] Cf. Rotter, 1980, p. 1 ff.
[33] Cf. Luhmann, 2000, p. 31
[34] Cf. Luhmann, 2000, p. 96
[35] Cf. Luhmann, 2000, p. 97
[36] Cf. Luhmann, 2000, p. 97

System-inherent mechanisms are therefore set up to help delay the reaching of the moment of distrust and to prevent it from becoming a destructive process for the system.[37] These can be measures such as general forms of transparent portrayal of errors, subsequent explanations, status declarations, or even apologies. It is, however, to be considered that once the client has been disappointed, these measures only serve to delay the threshold point, and not to completely eliminate it.

2.1.3.3 Interpersonal and Systemic Trust

Luhmann furthermore distinguishes two forms of trust: interpersonal and systemic trust. Interpersonal trust is created during the interaction of two individuals. Its creation and growth takes time.[38] It is built in little steps in a reciprocal process of advance service and reward.[39] System trust is different from interpersonal trust. As a mobile payment service is similar to a technical system, the following paragraphs will cover systemic trust in greater detail.

Where Luhmann speaks of systems, he defines them as delineated from their environment. In this definition, environment is described as the indefinite shapes of possibilities and infinite potentials that surround a system in the real world.

---

[37] Cf. Luhmann, 2000, p. 100
[38] Cf. Bohn, 2007, p. 25 ff.
[39] Cf. Luhmann, 2000, p. 55

It therefore has a much higher complexity than the system, which delimits itself, and is described as "the total of updatable and non-updatable possibilities".[40]

The system manages to reduce high environmental complexity to a closed order. In Luhmann's words, a system is "organized complexity that operates through the selection of an order."[41] The term system has its origin in the ancient Greek word *sýstēma* and means a "whole composed of several parts".[42] This means systems are a composition of elements and relationships. Luhmann's view is that any social order can be described as a system of participants and their communication. Each family, party, community or enterprise is therefore a social system. Luhmann focuses on social systems, but does not discuss the term system with respect to the person/machine interface. Accordingly, there is a research gap as a result of a missing study of the relationship of trust in a person/machine relationship.

Rohpol, for instance, describes a technical system as a technical formation consisting of virtual and physical components.[43] The physical components are technical devices, such as a cell phone or a laptop. These technical artefacts wouldn't have any

---

[40] For Luhmann quote, see Krause, 2005, p. 10
[41] For Luhmann quote, see Berghaus, 2003, p. 36
[42] Cf. Pfeifer, 1993, keyword "system"
[43] For systems theory of technology, cf. Ropohl, 1999

social legitimacy without the client's user trust. The virtual components are digital services, which only become usable through the physical device. The same is true here: There is no basis of legitimacy without the user's trust.

In contrast to interpersonal trust, system trust is related to the relationship between a person and a (technical) system. Interpersonal trust requires communicative interaction between people in order to build credibility.[44] Verbal and non-verbal communication triggers a feedback process, which enables the creation of familiarity and trust. This interpersonal empathy cannot be performed by a technical system. The missing interaction frequently leads to a unilateral initiation of communication by the user, with the consequence that system trust is built more slowly than interpersonal trust. At the same time, it is achieved more quickly yet superficially and shows more resistance to disappointment.[45]

At this point, Giddens distinguishes between face-dependent and face-independent trust relationships.[46] A face-dependent trust relationship is rooted in the idea of the co-presence of two individuals and the building of a social connection based on mutual credibility. The face-independent

---

[44] Cf. Kumbruck, 2000, p. 109
[45] Cf. Luhmann, 2000, p. 75
[46] Cf. Giddens, 1995a, p. 20

trust relationship between a human being and a machine, on the other hand, does not come about through the co-presence of two individuals, but through the compliance with certain obligations and rules. In the case of mobile payment services, these rules refer to aspects such as usage procedures, functionality, systemic feedback, reliability, and security. They serve the organization of the relationship and the measurability of trustworthiness. The system only acts outside of regular programming standards when it comes to resolving errors. Learned symbols, standards, and rules thus gain significant signal effect to produce inner security with the user.[47] Fukuyama states that standards and rules promote system trust, while they weaken interpersonal trust.[48]

Lewicki & Bunker even describe system trust as generalized trust.[49] The use of a system by others generates inner security in the user. Thus, system trust is based on the imitation of other people's selections.[50] The exact control of a mobile payment service would require a high degree of expert knowledge. As in the principal-agent dilemma, the expert has a knowledge that gives him an edge over the user and that he can employ in the latter's

---

[47] Cf. Luhmann, 2000, p. 62
[48] Cf. Fukuyama, 1995, p. 5 ff.
[49] Cf. Lewicki/Bunker, 1995a, p. 137
[50] For the concept of "*institution-based trust*" cf. Zucker (1986); for the concept of "*third party trust*" cf. Coleman (1982) – for "*trust and third party gossip*" cf. Burt/Knez (1996); for fundamental concepts of system trust cf. Giddens 1995a.

favour or to his detriment.[51] The user doesn't have another choice than to leave control in the expert's hands, which supervises the system. This means that trusting the technical service always also means trusting the people behind it.[52] "Individuals must generalize their personal trust for large organizations made up of individuals with whom they have low familiarity, low interdependence, and low continuity of interaction."[53]

System trust doesn't have to be learned from scratch every time. It even has proven particularly resistant to disappointments. Disappointments with systems have the rare quality that they are not an interpersonal issue, but an individual one, and thus hardly get any publicity.[54]

However, system trust cannot be completely separated from interpersonal trust, since the organizations or technical systems are operated and represented by people, after all. Even in the case of technical systems, a personal aspect of trust building must always be taken into account, especially in the case of personal client contact or the trust-building impact of representatives of a mobile service.[55] The significance of the transmission of convictions, values, and rules by representatives that creates

---

[51] Cf. Mathissen, 2009, p. 17 ff; Luhmann, 2000, p. 77
[52] Cf. Sztompka, 1999, p. 21, 41 ff
[53] Cf. Lewicki/Bunker, 1995, p. 137
[54] Cf. Luhmann, 2000, p. 78
[55] Cf. Rupf-Schreiber, 2006, p. 101

meaning and relationships can be assessed as extremely important in the development of a culture of trust.

2.1.4 Determinants of Trust

In her book 'Trust in Modern Societies' Barbara Misztal undertakes the comprehensive attempt to analyse trust in modern societies by studying traditional works on trust in the area of social sciences, by authors like Weber, Simmel, Spencer, Durkheim, and Toennis, and developing her own approach.

2.1.4.1 Classification

Mizstal is to a certain degree relevant in this context as she presents an overall analysis of the concept of trust and defines three types of trust in a modern society. These are Habitus, Passion, and Policy.

She allocates a social order to each of these manifestations.[56] "... three types of order should be distinguished: *stable order*, which accounts for the predictability, reliability and legibility of the social reality; *cohesive order*, which can be seen as based on normative integration; and *collaborative order*, which refers to social cooperation."[57]

However, in her classification, she does not only attribute a social order to each type of trust, but also

---

[56] Cf. Misztal, 1996, p. 101
[57] Misztal, 1996, p. 64

formulates initial practical indicators of trust, which are different for each type of trust.[58] Figure 1 shows the classification of types of trust according to Misztal.

Order	Trust	Practice
Stable	Habitus	Habit, Reputation, Memory
Cohesive	Passion	Family, Friends, Society
Collaborative	Policy	Solidarity, Toleration, Legitimacy

Fig. 1 – Types of trust according to Misztal

If a social relationship has steady, stable character, trust corresponds to a primarily behaviour-based relationship, which can be generated and influenced by habit, reputation, and memory. In cohesive relationships, Misztal describes trust as a passionate tie, which is often created through a common value system or goal, as can be found in families, friendships or social communities of interest. In a collaborative relationship, trust has the character of a methodical, regulated connection and is influenced by solidarity, toleration, and legitimacy. In this book, it is assumed that the behaviour-based relationship according to Misztal corresponds to the

---

[58] Cf. Misztal, 1996, p. 101

trust relationship between an individual and a mobile payment system. The user's behavioural rules organize the human/machine relationship, while technical stability and security are the main columns for user trust. Habit and reputation are therefore the central determinants of the behaviour-based trust relationship. Both will be covered in more depth in the next chapter.

2.1.4.2 Habitus

Pierre Bourdieu coined the term *habitus*. He describes habitus as follows: "What I call *habitus*, acts as a connecting link between the position or rank within a social space and specific practices, preferences, etc.; it is a general, fundamental attitude, a disposition towards the world, which leads to systematic statements. There is...in fact...a connection between highly disparate things: how someone talks, dances, laughs, reads, what he reads, what he likes, what acquaintances and friends he has, etc., is all closely intertwined."[59]

In other words, the habitus is an agent's concept of social interaction that reflects his attitude, disposition, value system, view or lifestyle. Bourdieu related the term mostly to individuals of the social world.

If the term habitus were used for a behavioural pattern of a technical system, this would mean that

---
[59] Cf. Bourdieu, 1992, p. 31

a mobile payment service is trusted because of its trustworthy or learned interaction behaviour. An essential element of the perceptibility of a technical habitus is basically the design of the human/machine interface.[60] Learned process operations and interactions help build trust on the basis of familiarity and understanding. Here, the perceived screen design can reflect the service provider's attitude with regards to performance features, such as qualification, integrity, or competence.

Furthermore, the trust relationship with a mobile payment service is strongly influenced by its reputation. Especially in times of social, interactive media that lead to enhanced transparency, the reputation of any agent in a society becomes valuable social currency. Due to the high relevance of reputation in an increasingly networked society, it will be discussed in the next chapter.

2.1.4.3 Reputation

In her book, Misztal puts special focus on the phenomenon of reputation.[61] She examines reputation mainly in the context of economic relationships. She points out that human beings prefer or neglect other people due to preconceptions, prestige, or clichés. The process of evaluating and verifying the credibility of a social

---

[60] For design of interactive interfaces cf. Tidwell, 2010
[61] Cf. Misztal, 1996, p. 120 ff

object is to a high degree based on their reputation, according to Misztal.

"A good reputation allows economic agents locked into the relation to cut the transaction costs and overcome limited information, and thus to facilitate efficient contractual relations. A good reputation in all types of business attracts customers and clients and increases the company's competitive advantages."[62]

Reputation has three mechanisms: moral codex (values), social adaptation (reciprocity), and formal control (sanctions, observation, discipline). These mechanisms don't possess exclusivity, but are all equally verified and evaluated by the observer.

Especially in highly technologized societies with digital media that permeate day-to-day life, a good reputation seems to rank very high when it comes to trust evaluation. The market participants' trusting willingness to use a service reduces the complexity raised by digital media with the effect of bridging insecurities and accelerating decision processes. The evaluation and recommendation of a service produces a trust-promoting discourse around it. In this way, its reputation in the social space is created. Improving and maintaining one's positive reputation becomes more and more challenging in a world that is digitally networked at all times.

---

[62] Misztal, 1996, p. 121

In conclusion, it should be pointed out that it is impossible to present the full scope of Misztal's theory here: She dissected the term trust much more comprehensively, showing a highly multifaceted image of the term. Habitus and reputation are central aspects of trust, but only two of many.

2.1.5 Creating Trust

After introducing Luhmann's functions and Misztal's determinants of trust, we will now see in which stages trust is created. Moreover, the role of trust within the framework of the economic decision-making process of the Technology Acceptance Model and of information technology will be discussed.

2.1.5.1 Phases of Trust

While Luhmann's approaches cover the nature of trust, and Misztal's approaches the determinants of trust, Lewicki & Bunker deliver insights into the developing stages of trust building. In their approach, they distinguish three stages of trust building.[63] (Cf. fig. 2.)

---

[63] Cf. Lewicki & Bunker, 1995b, p. 114 ff.

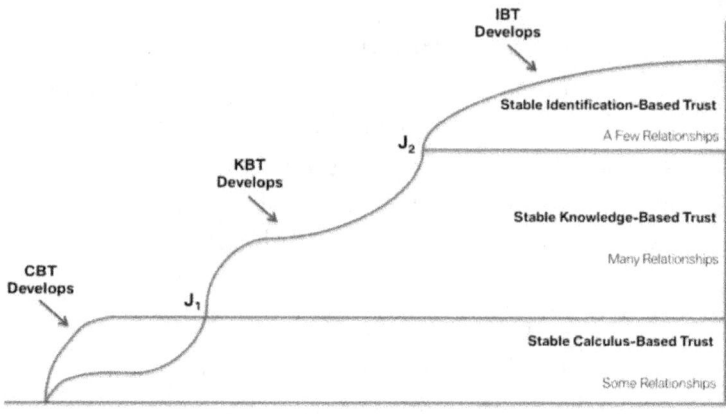

J_1: At this point, some calculation-based trust relationships become knowledge-based relationships.
J_2: At this juncture, a few knowledge-based trust relationships were positive affected and go on to become identification-based trust relationships.

Fig. 2 – Stages of trust building according to Lewicki & Bunker

The first stage has a situational character and is based on the trust giver's calculations. Lewicki & Bunker call this highly rational form of trust *Calculus-Based Trust* (CBT). This form is mainly created between social agents who basically cannot support their trust with experiences and therefore have to apply economic tactics. This stage of trust can be related to the principle of utility maximization as well as the Rational Choice Theory.[64] Due to a lack of familiarity, knowledge, and common experiences, the trust giver does not have a choice but to make assumptions with regards

---

[64] The Rational Choice Theory is an approach to the action theory in economics and social sciences. It assumes rationally acting agents in social constructs, which pursue the objective of utility maximization. For Rational Choice Theory cf. Diekmann/Voss, 2004, p. 13 ff.

to his counterpart's trustworthiness with the help of symbols, characteristics, and reputation (cf. reputation Misztal). During this phase, the giving of trust is characterized by a high degree of insecurity. Like in a strategy game, possible risks are balanced against potential gains. Trust will only be granted if the usage can potentially be maximized in the future.

With repeated interaction during the next phase, empirical knowledge is created between the parties. The prompt willingness to trust at the moment of taking a decision increases with more positive experiences in the past. Lewicki & Bunker call this next phase of trust *Knowledge-Based Trust* (KBT). The willingness to trust is now no longer rooted in an economic calculation, but in personal experiences. Because of its nature, this form of trust can only be developed through a certain expenditure of time and communication.

Lewicki & Bunker call the third phase of trust *Identification-Based Trust* (IBT). Assuming that the trust granted is not disappointed over an extended period of time, repeated interaction and common goals and experiences lead to an emotional tie. In this phase, the trust giver reaches the highest willingness to trust the trust taker. The long interaction history enables a strong connection and even leads to the agent's identification with the object of trust. In this phase of trust, there is hardly any scepticism or critical attitude left at the moment of decision-making.

The IBT stage can, however, only be built over a long period of time with a high number of emotional interactions.[65] But it needs to be considered that the scope of interactions will likely be limited, as system trust relies on a strict set of procedures. (Cf. 2.1.3.3 Interpersonal and Systemic Trust.) Especially in the case of technical systems, a spontaneous, irregular interaction would possibly rather lead to mistrust than to the promotion of trust on part of the user. At the same time, however, one needs to ask whether this is different in the case of mobile services because of their nature as an intimate, personal communication device. There is possibly a form of communication between the mobile service and the user that leads to a closer connection and to a higher intensity of trust.

2.1.5.2 The Economic Decision-Making Process

Rational Choice Theory offers a relevant approach in examining the term trust in the context of economics.

The analysis of economic activity has a longstanding tradition in economics. The Rational Choice Theory

---

[65] Cf. Lewicki/Bunker, 1995b, p. 114 ff.

explicitly studies the behaviour of agents at the moment of decision-making. It assumes that the basis for an agent's decision is always maximization of utility. This consideration is based on the image of the human being as the Homo oeconomicus: a person whose decisions are always rational, selfish, and profit-oriented.

This image has repeatedly been criticized as unrealistic, as it ignores many human facets, including the moment of trust in the exchange of resources. In economic transaction, trust is an elementary prerequisite for an exchange relationship.[66]

The agents of an exchange relationship are always faced with the insecurity of never having all the information necessary to make a perfectly rational decision. In fact, they only receive a fraction of decision-relevant criteria for their assessment of the situation. Besides the content-related asymmetry, there is often also a time-related one. In the case of transactions, the trust giver must provide resources in advance in order to receive his reward. Here, the most important function of trust is to overcome economic risk and/or insecurities.

On this basis Coleman developed the economic trust concept.[67] In order to make a decision to act,

---

[66] Cf. Coleman, 1991, p. 40, p. 45 ff.
[67] Cf. Coleman, 1991, p. 45 ff.

the trust giver evaluates the trustworthiness of the situation. A positive decision depends on whether the chances of winning are bigger than the chances of losing.

The decision-making moment can be expressed as the formula in Figure 3.

---

p = chance of winning
(Probability of the trustee's trustworthiness)

L = potential loss
(If trustee is not trustworthy)

G = potential gain
(If trustee is trustworthy)

Decision

Yes if $\quad p/1-p >\quad L/G\quad$ and/or $\quad pG > (1-p)L$

Undecided if $\quad p/1-p =\quad L/G\quad$ and/or $\quad pG = (1-p)L$

No if $\quad p/1-p <\quad L/G\quad$ and/or $\quad pG < (1-p)L$

Fig. 3 – Formalization of granting trust according to Coleman

The goal of this economic assessment is to rate trustworthiness, which results in a positive or negative trust attitude. The agent considers the potential loss that he would suffer if the trust taker abuses his trust and the potential gain he would obtain if his trust were justified, as well as his chance to win, defined as the probability of the trust taker's trustworthiness.

While the potential loss L and the potential gain G can be named with relative precision, an interpretation of the assessment of the probability p is incomparably more complex.[68] A high probability indicates trust - a low probability distrust.[69] Hence in some way the process of determining the probability becomes the phase of information search, as p is a result of the maximum information available.[70] The agent perceives the relevant information in a subjective way: on the one hand, based on inner, learned attitudes and preferences, and on the other hand, as a result of the social context, such as external, social norms and realities. Sydow describes trust in this context as a concrete expectation, which is always connected to a specific decision-making situation.[71] The trust giver collects the maximum possible information as to incentives, restrictions, or appropriateness in order to determine his probability to win.

If a user evaluates the trustworthiness of a mobile payment service along this model, the following data can be valuable parameters for his assessment:

---

[68] Cf. Coleman, 1991, p. 129
[69] For the evaluation of trust cf. Gambetta, 2001, p. 211
[70] Cf. Coleman, 1991, p. 131
[71] Cf. Sydow, 1998, p. 35

p: Relevant information for the probability is data related to standards, reputation (providers, merchants), familiarity, clarity (compatibility), moments of control, ease of use (availability of the components of the service), new or learned alternatives for action, authentication process, and security features.
L: Amount to be paid, time (to resolve fraud), data.
G: Availability of the merchandise purchased, possible time and money saving.

2.1.5.3 Technology Acceptance Model

Approaches to adoption, diffusion, and acceptance research are useful in finding out how consumers accept innovations and what factors have a significant impact on how widespread it becomes.

When considering trust with regards to the provision of a novel mobile payment service, adaptation research is particularly helpful. The positive adoption of a technical innovation can be understood as evidence of a willingness to trust. In his study, Bornschier states that trust is an essential factor when overcoming insecurity at the moment of adopting a new technical solution.[72] "According to our results, the cultural resource of generalized trust...has a proven impact on the early diffusion of the new Internet technology," says Bornschier.[73]

---

[72] Cf. Bornschier, 2001, p. 237
[73] Bornschier, 2001, p. 253

Within the framework of information system research, Davis delivered the basis for the *Technology Acceptance Model* (TAM) with his research of user acceptance in 1989.[74] (Cf. fig. 4.) The model states that the two main factors Perceived Usefulness and Perceived Ease of Use significantly influence whether clients actually use a technology or not.

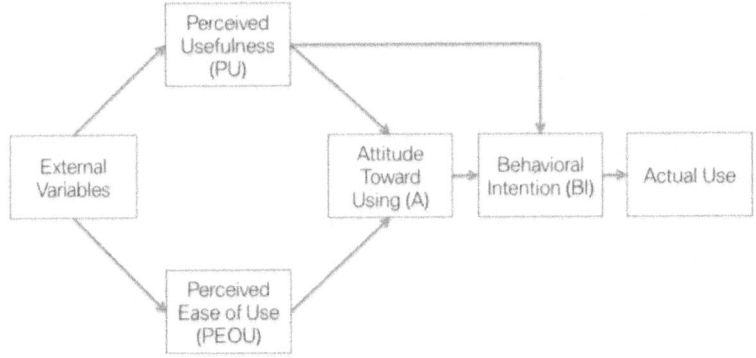

Fig. 4 – Technology Acceptance Model (TAM) according to Davis

Both the *Perceived Usefulness* (PU) and the *Perceived Ease of Use* (PEOU) determine the *Attitude Toward Using* (A) the new technical solution. The *Behavioural Intention* (BI) is influenced by PU and A. And BI eventually triggers the actual use.

However, the Technology Acceptance Model is not applicable to mobile services, as the nature of mobile services entails a much more intimate and closer interaction than that of stationary services.

---

[74] Cf. Davis/Bagozzi/Warshaw, 1989, p. 985

Based on the original model, Kaasinen formulated two additional characteristics that have an influence on the acceptance of a new mobile service.[75] (Cf. fig. 5.) These are Trust and Ease of Adoption. Kaasinen thinks that the factor Trust already becomes important for the user in an early phase of the decision, namely in the initial reply to the intention to use.

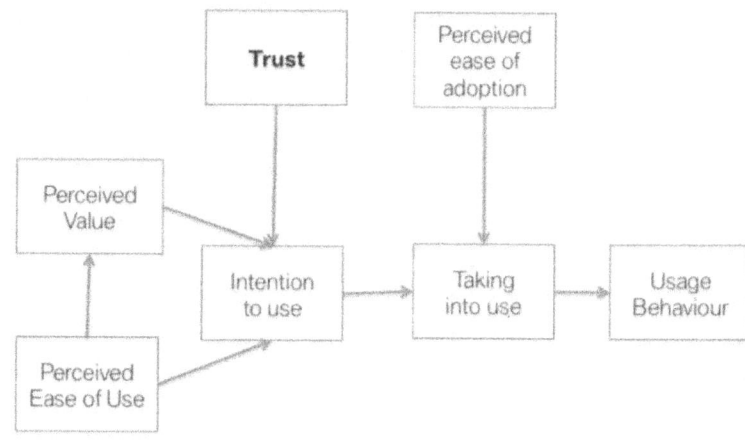

Fig. 5 – Technology Acceptance Model for mobile services according to Kaasinen

In their studies of trust, scientists like Neumann or Thalmann connect this trust to aspects such as competence, quality of communication, or presentation of information.[76] In this context, it is important to ask whether the contents of Perceived

---

[75] Cf. Kaasinen, 2005, p. 72
[76] For further aspects of trust cf. Neumann, 2007 and Thalmann, 2005

Value and Perceived Ease of Use already overlap significantly with those of presentation of information, quality of communication, or competence?

In this case, trust would not be a separate factor, but the factor – shaped and influenced by the Perceived Value and Perceived Ease of Use aspects – that precedes the step Intention to Use. Perceived Value and Perceived Ease of Use thus become the initial relevant factors of influence to secure the user's willingness to trust. And so the trust factor becomes the only preliminary stage of Intention to Use.

Perceived Value and Perceived Ease of Use illustrate the high significance of both the communication model and design as central tools to promote trust. Kaasinen also recognizes the high relevance of design for the acceptance of mobile services.[77] However, he does not go into detail about the classification or creation of specific determinants and elements for trust creation.

2.1.5.4 Tech Trust as Qualified Reliability

The complexity of modern technical systems makes it difficult for users to quickly grasp digital products. The widely ramified systems with their non-transparent structure and processes are often nearly impenetrable for users.

---

[77] Cf. Kaasinen, 2005, p. 70

Weber stated that due to the impossibility of figuring out the developing and operational processes of technical products, clients must concentrate on the important expectations with respect to the practical behaviour of a technical artefact.[78] Indirectly this means that complex technical systems force the user to be naively open and that it becomes normal to blindly trust digital or mobile services. Latour says: "Rely on machines is balanced by the counter-advice of never trust them."[79] He refers to the eternal insecurity pertaining to technical systems.

In 1997, Gerck studied the function and form of trust with regards to digital systems. The information theory is his theoretical frame of reference, and the Internet his practical field of application. His thesis is that with the provision of trust in the digital world, digital communication becomes more similar to person-to-person communication.[80] Digital communication still misses one dimension of meaning. In real communication, this dimension is created through an emotional or physical experience. Digital communication, though, is limited in its interaction elements, as the nature of digital communication primarily serves the transfer

---

[78] Cf. Weber, 1988, p. 465
[79] Latour, 1988, p. 29
[80] Cf. Gerck, 2002, p. 20 ff

of interpreted and related content packages or Internet protocols.

Therefore, Gerck does not conceive trust in digital space merely as a synonym for faith, confidence or ingenuous openness. In digital space, trust has the central task of qualifying information as to reliability, according to Gerck. As a result, trust becomes a reliability feature for information received.

As trust in the digital world does not have anything to do with emotions or feelings, trust in the digital TCP/IP world between human being and machine must also be communicable or transferable. In this connection, Gerck asks for transfer rules because a unilateral statement of the accuracy of digital information is no longer enough for the creation of trust in a digital context. One-dimensional client-server communication is also insufficient to build trust. Gerck suggests open control mechanisms in the network model, which exceed two communication units. Multidimensional networks with units that communicate independently seem more appropriate for open information control. Gerck does not suggest concrete control models for trust in the digital context.

Nonetheless, his deliberations have special relevance with regard to the applied description of the trust aspect in the digital relationship of tension between a human being and a mobile payment service. These control mechanisms play a central role in the design of payment processes. The creation of the process steps and feedback elements are trust-sensitive elements of a mobile payment process. The payment process will be covered in more detail in 2.2.4. The challenge lies in the confidence-inspiring design of a service in the area of tension between the often very contradictory elements of security and user-friendliness.

2.1.6 Summary

Trust is a fundamental element for overcoming risks in social interactions within a society. In the case of payment processes, it is considered the justification of their existence. Without trust, payment media and their payment processes are not used. Especially in the case of technical systems, trust serves to reduce complexity, to overcome risk or to positively anticipate future events. Compared to interpersonal trust, system trust is hardly perceived in day-to-day life. While the granting of personal trust implies a conscious evaluation, system trust rather is based on generalization and subjective familiarity. Trust and distrust only become a conscious issue when the system does not work, as it should according to its rules. The concrete trust situation between a technical system and a human being has not yet been analysed in detail in the literature. Neither Luhmann nor Latour provide a concrete answer about its function or aspects. Although Luhmann

studies system trust, he does not relate it to technical systems; Latour does cover the system theory as related to technical systems, but without including the topic of trust.
In order to develop and intensify trust, regular positive experiences and interactions are necessary.

On the one hand, these are based on the functioning of the rules agreed on, and on the other, on the appropriate reaction whenever system errors occur in order to avoid crossing the threshold of distrust. In this context, several different aspects of trust can contribute to increasing trust and avoiding distrust. Aspects such as habit, reputation, and probability to win were previously considered. Some other scientists describe aspects of trust as shown in the overview of Figure 6.[81]

Aspects of Trust	Author
Sense of responsibility, readiness to help, goodwill, competence, consistency, honesty, fairness	Morgan/Hunt, 1994
Competence, honesty, openness, effort	Laucken, 2005
Confidentiality, expert competence, reliability, fairness, quality of communication	Büssing/Moranz, 2003

[81] The overview does not lay claim to completeness.

Competence, predictability, goodwill, integrity, honesty	Einwiller, 2003
Tolerance, discretion, loyalty, integrity, honesty, reliability, openness	Picot/Reichwald/Wigand, 2003
Appreciation, patience, quality of experience	Helm/Gehrer, 2006

Fig. 6 – author's overview, aspects of trust (author's illustration)

Aspects such as competence, reliability, or discretion are cited in this context. The authors also mention openness, honesty, and quality of experience as aspects of trust. Here it becomes obvious that the aspects of trust take on a different meaning depending on the object. In relation to mobile payment processes, these aspects could be security, transparency, and user experience. Transparent communication, technical security or generally intelligible user interfaces would therefore be trust-promoting factors. The section Practical Comparison examines which aspects of trust are regarded as relevant in practice, based on interviews with representatives of the industry.

2.2 About Mobile Payment

In the past, there were many forms of payment media. Shells, cowries or gold coins were used as means of value transfer. In the meantime, the transfer medium has evolved from a bank bill to digitalized payment data via Internet or mobile phone. With the expansion of e-commerce, the need for a digital payment process has further increased. Without any real contact between buyer and seller, the payment data must be transmitted in a fast and secure manner through an Internet connection. With the transformation from a physical to a digital payment process we are now experiencing a change of paradigm in the transfer of values within our society.

Not the possession of physical money, but the permanent availability of its stored assets is relevant. Money becomes invisible and disappears into the digital payment process, or dissolves in a mobile phone. The diffusion of mobile data plans, fast data connections, and high-tech smartphones opened up another channel for the virtual transfer of values: the mobile terminal.

The mobile phone, which has advanced to the position of the seventh mass medium, has on a global level reached an average penetration rate of 86.7% as measured by active mobile phone contracts.[82] This means almost 6 billion people have a wireless device.[83] With so many people using wireless devices, they have become popular for mobile payment services. Furthermore, the expansion of smartphones and mobile applications has led to more sophisticated technical options and improved user interfaces for mobile payment services.

Many analysts predict enormous market growth for mobile payment. According to Juniper Research, by 2015 the transfer volume of mobile payment processes will rise to $ 670 billion.[84] Mobile payment solutions have been implemented successfully, especially in developing countries. M-Pesa, for example, is now the largest driver of sales for Vodafone with 14 million clients in Kenya.[85] In Western Europe and the U.S., new solutions, such as Square, Google Wallet, Barclays Pingit or PayPal's Mobile Payment Application brought fresh air to the market of mobile payments. It is obvious that mobile payment is gaining more visibility from the large market players, even if in Europe and the U.S. it is still in an early stage of development.

However, the success factors of digital payment media are only marginally different from those of

---

[82] Cf. mobiThinking (February 2012): Global mobile statistics 2012
[83] Cf. mobiThinking (February 2012): Global mobile statistics 2012
[84] Cf. Rao (5 June 2011)
[85] Cf. mobile money l!ve (September 2011): M-Pesa the largest source of revenue for Vodafone in Kenya

physical payment media. They also must build the clients' trust through security, data protection, and user friendliness.

According to a present GfK (Gesellschaft für Konsumforschung – Society for Consumer Research) study, the trust factor represents a significant driver of the willingness to use new mobile payment services. The study was conducted in nine countries (USA, UK, France, Italy, Germany, Spain, South Korea, Brazil, and China) with the help of 8,000 online interviews.[86] The result was that consumers show a significantly higher willingness to adopt a mobile payment service when they trust the service provider or brand.

Based on the analysis of the results, the study describes the path of adopting a mobile payment service in three phases: trust, consideration, and preference. During the first phase, the trust strategy acts as a gate between user and service. Here, the factor trust has sort of a gatekeeper function. The granting of trust initiates phase 2, and the potential user is now ready to consider the new service. If this analysis also has a positive outcome, phase 3 follows. In this phase, the client starts employing the service, and if his experiences with it are mostly positive, he will give it priority over other services.

---

[86] **Cf. GfK Study** "*Mobile Payments: Trust and familiarity are crucial in driving adoption.*"

The following chapter will cover the peculiarities and forms of mobile payment in more detail. A basis will be created to better assess the need for trust by means of an in-depth study of the concept definition, the ecosystem, the classifications, the payment processes, and the stakeholders.

2.2.1 Terminology

The mobile payment market with its many solutions has been very fragmented to date. The most diverse technologies, value chains, form of cooperation, and business models have led to a wide variety of solutions. Therefore, the term mobile payment will be more specifically defined and narrowed down in the following chapter

2.2.1.1 Definition

If we orient ourselves along the definition of the term payment by the European Central Bank, mobile payment describes the transfer of a monetary claim triggered by a sender or accepted by a receiver by means of a wireless device. Thus mobile payment (m-payment) can be described as a process between two agents that serves the monetary value transfer via a wireless device, such as a feature phone, smartphone, tablet computer, or personal digital assistant (PDA)

2.2.1.2 Delineation

The term mobile payment is often used with relation to mobile money, mobile banking, and mobile commerce. These terms overlap in their areas of application, but they are to be clearly differentiated according to the definition by the Global System of Mobile Communication Association (GSMA).[87]

Mobile money is a purely monetary value unit, which is usually stored in a bank account and employed or retrieved via a wireless device. In contrast, mobile banking describes the administration or access of a bank account by means of a wireless device. Hence, mobile banking allows mobile account administration, mobile portfolio management, and access to mobile financial data. Mobile commerce comprises the ability to trade via a wireless device. The term includes any trade that implies the transfer of rights and obligations related to the exchange of goods and services triggered and/or completed by means of a wireless device.

2.2.2 Evolution of the Ecosystem

At the beginning of the 1990s, the mobile phone obtained its first relevant visibility due to the introduction of the digital D-network.[88] As a

---

[87] Cf. GSMA (July 2010): Mobile Money for the Unbanked. Mobile Money Definitions
[88] Cf. ITU (December): Mobile cellular subscriptions. Key 2000-2010

portable phone, it allowed for location-independent communication by radio signal via a telephone network. While the devices served mainly purposes of voice telephony, they were complemented with services such as SMS, MMS, USSD or mobile Internet, and thus became so-called feature phones. But the mobile ecosystem evolved further, as did users' expectations.

2.2.2.1 Transmission Technology

Increasingly efficient and secure transmission technologies as well as decreasing rates of data plans provided the conditions for wireless devices to be seriously considered for use as a payment medium.

With diverse technical interim stages, mobile transmission technology evolved from the GSM (Global System for Mobile Communications) and the UMTS (Universal Mobile Telecommunications System) to the LTE (Long-Term Evolution) standard.

GSM became the global mobile radio standard of the so-called second generation. It was introduced in 1992 as the technical basis for the D-network. At the time, mobile Internet didn't play a significant role yet as transfer tariffs were still poorly conceived and connection costs high.

The GSM standard reached a maximum data transmission rate of 9.6 Kbit/s. For voice transmission, this was sufficient, but for data transmission it was achingly slow. With the expansion of the GSM standard by the (High Speed Circuit Switched Data), GPRS (General Packet Radio Service), and EDGE (Enhanced Data Rates for Global Evolution) standards, a maximum transmission rate of 470 Kbit/s was achieved.[89]

In 2000, the UMTS standard followed as the third generation. The spectacular license auction of the Bundesnetzagentur (Federal Network Agency) was in the headlines in Germany.[90] The technical and economic expectations for UMTS were accordingly high. The main difference to the GSM standard was the frequency bandwidth applied. While the second generation used a frequency range of 200kHz, UMTS used a range of 5 MHz, which is 25 times more extensive. Moreover, HSPA+ allowed for a downstream speed of up to 337.5 Mbit/s.[91] This enabled the transmission of multimedia services with voice, text, images, and videos.

In 2009, the fourth generation, LTE, was first launched in Norway and Sweden.[92] The UMTS successor LTE achieved considerably higher

---

[89] Cf. Wiecker, 2002, p. 435
[90] Cf. Federal Network Agency (18 August 2000): UMTS auction procedure concluded – with total amount of DM99.3682bn
[91] Cf.. Elektronik Kompendium: HSPA
[92] Cf. Ali-Yahiya, 2011, p. 10

download and transmission rates than its predecessor. With LTE-Advanced, transmission rates of up to 1 Gbit/s were possible.

With the UMTS standard, data transmission was packaged-based for the first time. Thus UMTS and LTE offer a considerably higher security standard than the preceding generations. The UMTS security architecture comprises the areas network access security, security in the network area, security in the user area, security in the application area, as well as visibility and configuration of the security mechanisms.[93] It is also the basis for the LTE security architecture. The main security mechanisms include the confidentiality of user data, authentication, the confidentiality of communication contents, data integrity, and the identification of the devices. Special security algorithms, authentication, and coding, offer the necessary confidentiality for the transmission of sensitive payment information.

However, despite the improved transmission technology, the general use of mobile data services or mobile Internet failed to materialize, as for a long time their use still involved high costs for end users. In Germany, this only changed as of 2006 through drastic price wars initiated by discount providers like Base. Data rates sank by up to 97% in terms of

---

[93] Cf. Reppekus (2000): Sicherheitskonzepte des UMTS Standards (Security Concepts of the UMTS Standard)

cost.[94] Ever since, data rates have become lower and lower. In 2009 alone the data rates decreased by 24%.[95] The expansion of flat-rate data plans led to new trust through capped cost security.

2.2.2.2 Smartphones and Usability

Despite the shrinking mobile market, IDC only observed growth in the smartphone segment during the first quarter of 2009.[96] The growth trend continued so that in 2011 total worldwide sales amounted to 472m smartphones.[97] Compared to the previous year, this meant a 54.7% growth. With these sales figures, the smartphone segment reached a market share of an impressive 31% of the total market of mobile terminals. According to IDC, in the fourth quarter of 2011, the leading smartphone manufacturers were Apple, Samsung, Nokia, RIM, and HTC.[98] For 2012, JP Morgan expected smartphone sales of 657m units.[99] With the rising sales figures, the innovation cycles of the terminal manufacturers were shortened. At the same time, the performance of smartphones was continuously increasing, which showed in the colour depth of the

---

[94] Cf. Schäfer, 2008, p.1, 4
[95] Cf. Go Smart: 2012 always-in-touch, Study of smartphone use 2012, p. 4
[96] Cf. IDC (30.06.09): Smartphone Growth Encouraging. Yet the Worldwide Mobile Phone Market Still Expected To Shrink in 2009
[97] Cf. Gartner (15.02.12): Gartner Says Worldwide Smartphone Sales Soared in Fourth Quarter of 2011 With 47 Percent Growth
[98] Cf. IDC (06.02.12): Smartphone Market Hits All - Time Quarterly High Due To Seasonal Strength and Wider Variety of Offerings
[99] Cf. Thomas et al. (16 Dec. 2011): Apple and Google in Christmas Showdown

displays and screen resolution. Especially the touch screen devices experienced high growth in the market.

The iPhone 4S with its A5 processor, for instance, reached a clock speed of 800MHz. This enabled more elegant interactions when opening and navigating applications. The also well-known Samsung Galaxy S2 with its dual core processor achieved a clock speed as high as 1.2GHz. This shortened the opening time for starting an application to less than a second.[100] The retina screen of the iPhone 4S with its resolution of 640x960 pixels attains a pixel density of 326 ppi and allows for sharper image display of the applications. In addition to higher processor performance, storage capacity grew as well. The iPhone 4S, for example, has an internal memory of up to 64GB. This created sufficient disk space for application-related data. Mobile operating systems such as iOS 5 also provide the option of storing encrypted data on the device.[101]

For mobile payment processes, there are now also technical approaches to integrate the secure element in a smartphone.[102] The secure element is a central

---

[100] Cf. Hillebrandt (23 May 2011): Bestes Smartphone 2011? Das Samsung Galaxy S2 im Smartphone-Test
[101] As to the options of data security with iOS 5 Apps cf. Warren, 2012, keyword "Data Protection"
[102] Cf. NFC Phones: Several hardware solutions for Secure Contactless Payments

security element in the payment process; protected auxiliary programs for the secure transmission of payment information are stored on it. One example of smartphone integration can be a physical chip such as a SIM card. Even Apple has patented the implementation of a SIM card as a secure element within the framework of a mobile payment NFC solution.[103]

In the fourth quarter of 2012, 50.9% of the smartphones sold globally already had the Android operating system and 23.8% the iOS operating system.[104] Both software manufacturers offered their own SDKs (software developer kits) to software developers. They contain a collection of auxiliary programs with user-friendly instructions for menu prompts and the design of interaction elements. Usability and service design gained new meaning with the iPhone and were again acknowledged as critical success factors.[105] Usability requirements experienced a boom, and the relevance of central usability criteria, such as effectiveness for the solution of a task, efficiency in the utilization of the system, and the satisfaction of software users, grew considerably.

---

[103] Cf. Free Patents Online (23.08.11): United States Patent Application US20110269423
[104] Cf. Gartner (15.02.12): Gartner Says Worldwide Smartphone Sales Soared in Fourth Quarter of 2011 With 47 Percent Growth
[105] *Usability* in German: *Nutzerfreundlichkeit*; in the sense of DIN EN ISO 9241 – Ergonomie Mensch System Interaktion (Ergonomics of Interaction of Human Beings and Systems) cf. Hartmann, 2008, p. 15 ff.

Mobile applications in particular, which are displayed on a small screen, require practicability, simplicity, and high speed. Since they are often used during breaks, when waiting in line, or when one is out and about, high usage intensity of the device can be achieved only by means of precise menu prompts, meaningful interaction elements and clear information architectures.

Given an increasing expansion of mobile payment applications in commerce and a more intense use by consumers, the 2015 projection for annual mobile payment transactions is $ 45 billion.[106]

2.2.2.3 Mobile App Boom

The introduction of app stores in 2008 brought a radical change in the usage of mobile services. App stores offered an ideal market place for software developers and end users. The number of applications provided as well as the enormous downloads figures led to an explosion of the market for mobile applications (apps). In 2011 worldwide app downloads were estimated at 18 bn.[107] This was an increase of 144% compared to the previous year. Apart from the Apple App Store and the Android Market, other market places were developed, e.g., Nokia Ovi Store, Blackberry App World, Samsung Apps, and Windows Marketplace.

---

[106] Cf. In-Stat (21 April 11): Annual Mobile Payment Transactions to Reach 45 Billion in 2015

[107] Cf. Chip Online (12 Sept. 11): App-Downloads: Android überholt Apple noch 2011

Mobile applications were rapidly adopted by financial institutions and third-party suppliers for a wide range of services. Some financial service providers even offered APIs (application programming interfaces) in order to provide the developers of m-commerce apps with integration instructions for their in-app payment solutions. PayPal was one of the first: In February 2010, it offered a programming library for the integration of a mobile express checkout.[108] This solution promised a fast and simple payment process for the existing PayPal user, as during the purchase, it was no longer necessary to enter payment data separately or to leave the application.

Most banks used mobile applications mainly for the implementation of mobile banking applications, and integrated the feature of mobile payment options. PayPal, for example, integrated the bump technology in their app as an additional function for claiming liabilities.

For several years, Visa has conducted a variety of pilots for mobile payment applications in order to test different application scenarios for the different markets.[109] Visa PayWave, a Near Field Communication- (NFC) based payment standard,

---

[108] Cf. PayPal (27 April 11): PayPal to Open Mobile Payments Library to Developers
[109] Cf. Visa: Visa Mobile Payment Pilots. Mobile Fact Sheet Chart

which is supposed to serve the data transmission between a wireless device and a point of sale (POS) payment terminal, was frequently employed in the pilot projects. In this context, Visa cooperated with banks and mobile providers in the corresponding countries. In the meantime, the number of third-party supplier solutions has increased on the market. Whether we look at software manufacturers such as Google or start-ups like Square: With their offers, the competitive and/or cooperative situation increases drastically while market complexity is enhanced. Start-ups also have the important market function of driving market activity with their innovative thrust.

The Twitter founder Jack Dorsey, for example, launched Square, in 2009. It allows any owner of an iPhone, iPad, or Android device to accept credit card payments. The target group of this service are small companies for which a traditional credit card reader would be too expensive or too complicated to use due to their mobile activity. The Square solution consists of a card reader dongle that transmits the payment data through the audio exit of the smartphone to the mobile Square app, and the Square app, which supports a standardized verification process. The payment data of the credit card is recognized by the application, the payable amount is keyed in, and the payment process is concluded by the client's manual signature. PayPal developed the competing product PayPal here.[110]

For the mobile wallet idea, lots of concepts have been developed as well. This idea considers the mobile application as a digital replacement of a physical wallet. The mobile wallet is meant to digitally replace plastic cards and cash and thus absorb certain functions of the wallet as well as accelerate the payment process at busy checkout registers. The currently best-known version is Google Wallet. It is an Android application, which allows the storage of the data of credit cards, loyalty cards, train tickets, flight tickets, and admission tickets, and to pay for or redeem them via NFC at a POS. For the data transfer between mobile device and terminal, the pilot project used the PayPass technology by MasterCard.

Trade chains like Starbucks are also provider of mobile payment applications. The Starbucks solution is called Starbucks Card Mobile App and is only accepted within Starbucks outlets. The mobile application is similar to a post-paid card, which can be charged by credit card or with a PayPal account for subsequent use as a payment medium at a Starbucks shop. An application-specific 2D barcode is used to debit the amounts payable. It is recognized and read by the cash register system.

---

[110] Cf. PayPal (15.03.12): PayPal Unveils PayPal Here: the First Global Mobile Payment Solution for Small Businesses

The value of the purchase is debited to the mobile card account after it has been registered.

With fast transmission rates, reliable network coverage, decreasing data tariffs, and improved usability, the basis for trustworthy mobile payment solutions has been established. But the most relevant question remained: will the consumer adopt the mobile payment solutions.

2.2.2.4 Market Acceptance

In this context, Capgemini Consulting conducted a comprehensive study of client perception with regards to mobile payment.[111] Overall, the study delivers insight into user behaviour, attitudes, and demographics, and it concludes that the users are now ready for mobile payment.[112] The online study shows that 60% of the respondents are familiar with mobile payment and the existence of some services. (Cf. fig. 7.) Most online users link mobile payment to transportation or travel-related applications, such as mobile parking (55.7%), mobile train tickets (43%), or mobile tickets for public transportation (37.3%).

---

[111] Cf. The online survey per online questionnaire, N=500, Capgemini Consulting, 2009, p. 1 ff.
[112] Cf. Capgemini Consulting, 2009, p. 11 f.

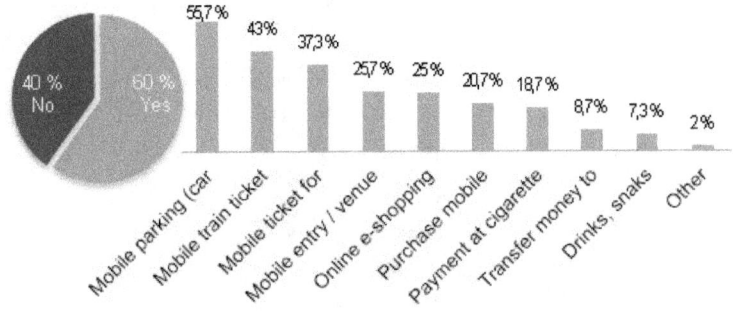

Fig. 7 – Familiarity with Mobile Payment offers according to Capgemini (Mobile Payment survey 2009)

At the same time, almost 45% of the respondents acknowledged mobile payment as an attractive service.[113] 37.6% of the questioned said attractive, and 7.1% said very attractive. (Cf. fig. 8.) The study measured attractiveness by means of a 5-point system, where 1 was not attractive at all and 5 was very attractive. The average value of attractiveness was 3.26 points. This group of highly interested people contrasts with another group of less interested users of 55.3%. The majority of the less interested individuals expressed a neutral point of view (36.9%).

---

[113] Cf. Capgemini Consulting, 2009, p. 11

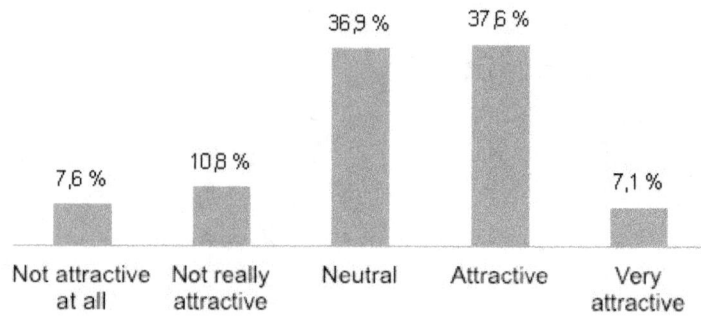

Fig. 8 – Attractiveness of mobile payment according to Capgemini (Mobile Payment survey 2009)

As the user still perceives the mobile payment service as a product that needs a lot of explanation and there is a positive correlation between attractiveness and familiarity, it is to be assumed, that in case of a neutral consumer group the value of attractiveness can simply be raised, by providing more information. Due to technical complexity, some mobile payment services may need to be more specific in their explanations in order to be accepted and preferred by the consumer.

Figure 9 shows which application scenarios are regarded as particularly relevant by the user.[114] A significant share of the respondents relates mobile payment systems with mobile commerce.
An overwhelming majority of 80.3% evaluate this payment scenario as highly relevant in the future. In contrast, acceptance in a consumer-to-consumer

---

[114] Cf. Khodawandi/Pousttchi/Wiedemann, 2003, p. 47

(C2C) scenario with 13.7% and in an electronic commerce scenario with 22.4% is remarkably low. Whether this means real rejection or just a lack of association is not clear from the study. Comparably high relevance can be observed with stationary applications in the interaction with vending machines (40.8%) or at store checkouts (27.1%).

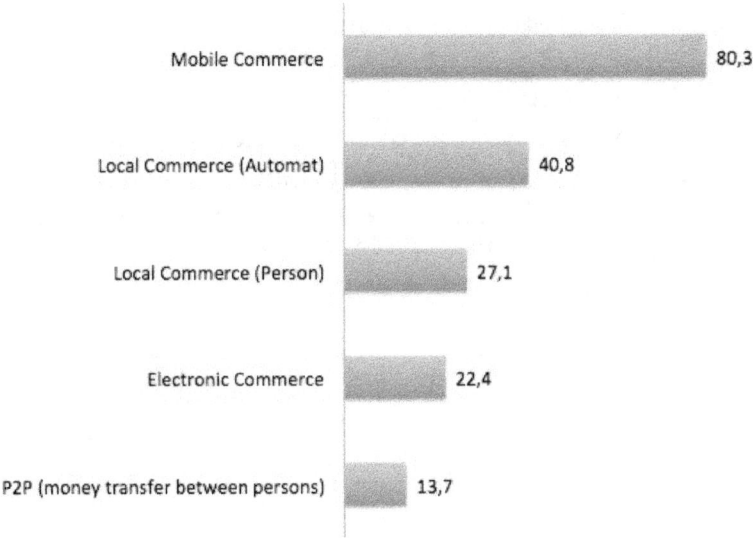

Fig. 9 – Relevance of payment scenarios according to Khodawandi et. al. (study of the acceptance of mobile payment procedures in Germany)

The results of Figure 9 show a correlation between existing points of acceptance for mobile payment and the personal assessment of their relevance. The subjective perception of relevance would therefore be strongly influenced by the actual penetration of points of acceptance. When asked for the reasons of rejecting a mobile payment service, the most frequently mentioned reason is subjectively perceived insecurity.[115]

The users' ideas about the general characteristics of a mobile payment service are also important. (Cf. fig. 10.) These expectations are possibly even indicators of central aspects of trust or trust-building features.

A high security standard, clear menu prompts, and a reliable performance of the service is indispensable, according to the users' evaluation. In comparison, a simple registration process and higher transaction speed are considered less important.

---

[115] Cf. Khodawandi/Pousttchi/Wiedemann, 2003, p. 46

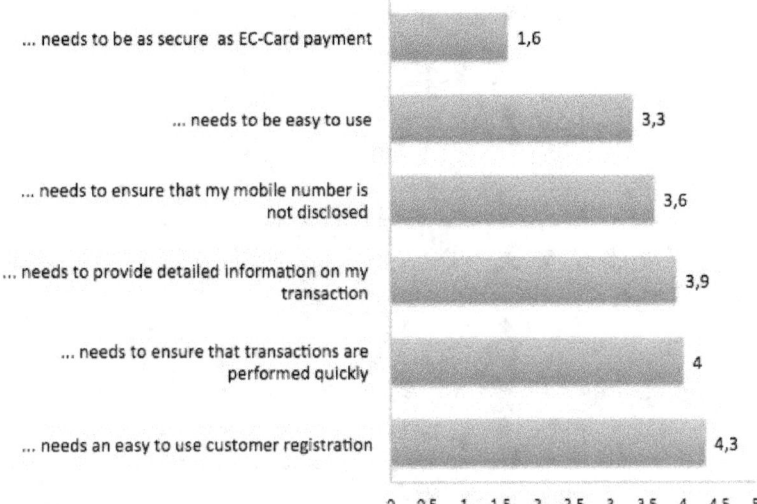

Fig. 10 – Important characteristics of mobile payment services according to Capgemini (Mobile Payment survey 2009). Rank 1 = highest, 6 = lowest

When examining individual criteria of acceptance, we can see a strong focus on security, performance, and transparency criteria. (Cf. fig. 11.)

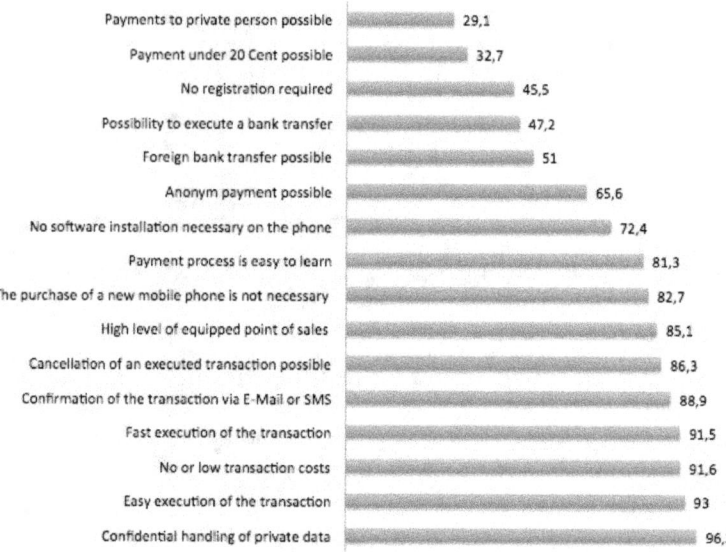

Fig. 11 – Characteristics of mobile payment processes according to Khodawandi et. al. (study of the acceptance of mobile payment procedures in Germany)

Whether it is the confidential treatment of personal data (96.2%), the simple handling of the payment process (93%), or the easy learnability of the payment process (81.3%), all of these items are indicative of a high significance of technical security, reliable performance, and user-friendly understandability.[116] Relevant trust aspects therefore can be technical certificates, the development of intelligent software, or user-oriented interaction design.

---

[116] Cf. Khodawandi/Pousttchi/Wiedemann, 2003, p. 48

2.2.3 Classification

The term mobile payment nowadays includes a large variety of technical systems, process structures, and applications. Since no unique, standardized application can be assigned to the term, below there is an attempt to categorize the manifold solutions. The different categories were defined through determinants such as place of participation, amount paid, and number of participating agents.

2.2.3.1 Place of Payment

The SEPA differentiates two types of mobile payment services: remote and proximity payments.[117] This distinction is based on the place of the payment process.

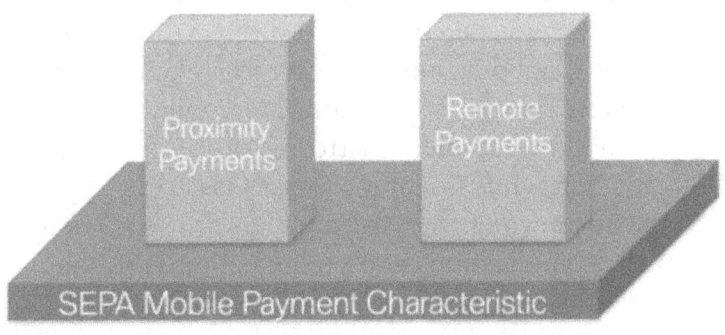

Fig. 12 - Mobile payment classification: Proximity and remote according to SEPA

---

[117] Cf. European Payments Council (07.02.12): Mobile Payments, p. 13 ff

In the case of a remote payment, the transaction takes place independently of the user's whereabouts. The payment path occurs through mobile Internet and corresponds to the acquisition of mobile services such as mobile tickets, the purchase of a mobile application, the charging of a pre-paid card, or the upgrade of a mobile game. It furthermore includes the purchase of a book via the mobile application of the Amazon online trading platform.

The technologies used are classical methods, such as Direct Operator Billing, Premium SMS, or Premium IVR. With the propagation of mobile data packages and the technical advancement of the mobile Internet from WAP to HTML5 protocols, it is now possible to integrate the payment elements of payment service providers and/or billing procedures via credit card or direct debit from classical online commerce on mobile optimized websites. Native applications are usually cleared through the payment platforms of the app store provider, while in-app purchases via APIs of specific payment service suppliers are integrated in mobile applications and processed through the payment provider.[118]

An example of a remote payment variant is Touch&Travel by Deutsche Bahn.[119] Via the mobile

---

[118] Cf. Google Developers: In-App Payments or Apple: Mac OS X Developer Library. Overview of In-App Purchase
[119] Cf. Touch&Travel

application, travellers can sign up for a trip before traveling and sign out after traveling. The ticket price is calculated based on the distance travelled and billed at the end of the month in accordance with the payment data stored.

Proximity payment, on the other hand, describes the transfer of payment data within the direct environment of the device. For transmission, technologies like Bluetooth, infrared, RFID/NFC or contactless smart chips are used. The technologies are implemented in the smartphone and in the POS payment terminal and allow for short-distance data transfer. Possible applications are cash registers and vending machines.

The Google Wallet service, for instance, uses this NFC technology at the moment a payment is transacted.[120] This application works in combination with an Android smartphone with NFC interface and a PayPass terminal at the POS. During the payment, the mobile phone is placed on the terminal. The user enters a PIN to confirm the amount before it is debited to the debit, pre-paid, or credit card that was activated for this use.

---

[120] Cf. Google Wallet. A description of the service can also be found in this work in the section on the stakeholders of the mobile payment ecosystem.

2.2.3.2 Payable Amount

Another distinctive feature of payment procedures is the amount payable of the transaction.[121] In finance, we distinguish between micro and macro payments.[122] A study on acceptance of mobile payment procedures in Germany states that there is high acceptance of mobile payment processes for any amount.[123]

Micro payment procedures describe transactions in which the payment values are minimum amounts of €0.10 to €5.00.[124] The services purchased are usually inexpensive products that do not require much explanation; so-called low-involvement products.[125] Mobile tickets, mobile applications, mobile games, news, or songs are well-known products in this category. Billing can be carried out through the clearing system of the service provider, direct billing of the mobile provider (MNO), or Premium SMS. The low need for explanation of the products should also be taken into account in the process design of the payment procedure. Mobile Payment should be fast and easy in order to be acknowledged as a serious alternative to traditional payment procedures.

---

[121] Cf. Teichmann/Nonnenmann/Henkel, 2001, p. 118
[122] Cf. Contius/Martignoni, n.y., p. 60
[123] Cf. Khodawandi/Pousttchi/Wiedemann, 2003, p. 53
[124] Cf. Kunde, 2001, S.7
[125] On the characteristics of low-involvement products cf. Kuß/Tomczak, 2004, p. 68

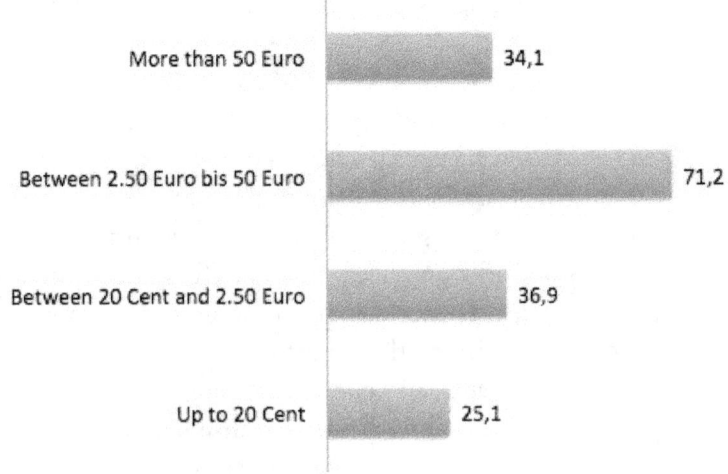

Fig. 13 – Amounts in mobile payments according to Khodawandi et. al. (study of the acceptance of mobile payment procedures in Germany)

Macro payment procedures start at a transaction value of €5.00.[126] Especially in the range of €2.50 to €50, there is a significantly high willingness to pay by mobile phone. (Cf. fig. 13.) Application scenarios in the offline world are particularly popular with clients. In physical commerce, mobile payment systems compete above all with credit and debit cards.

2.2.3.3 Time of Debit

The electronic billing procedure can also be classified by the point in time the debit takes place.

---

[126] Cf. Kunde, 2001, p. 7

Here we distinguish between pre-paid, post-paid, and pay-now.[127]

A payment made before delivery is called a pre-paid procedure. In this case, the client uses an account, which he credits with monetary units. He can only pay for goods or services with this account when there is a credit balance. PayPal offers such a pre-paid account, which can settle liabilities per pre-paid account as soon as the client charges it.[128] With the pre-paid method, the client makes an advance payment, while the supplier minimizes his credit risk. It is a mode of payment accepted by the customer for the settlement of small and minimal amounts. 34.8% of the respondents to a study by Khodawandi et. al., prefer to pay amounts of up to €2.50 by means of a balance-based mobile payment account.[129] Thus the pre-paid option ranks second highest in overall comparison.

If the goods or services purchased are billed at a later point in time, this is called a post-paid procedure. Billing occurs after a certain period of time has elapsed after delivery. In this case, the merchant or the provider of the payment medium assumes the risk of payment default. This means corresponding risk management must be provided for potential losses from fraud or bad debt. A

---

[127] Cf. Neumann, 2006, p. 116
[128] Cf. PayPal
[129] Cf. Khodawandi/Pousttchi/Wiedemann, 2003, p. 12

classical post-paid example is credit card payment. The client only receives a bill for all transactions at the end of the month. The amount owed then usually is settled through a debtor's account. The billing of the digital service of voice telephony by a mobile provider through a phone bill can also be assigned to the post-paid category. This is because the service is used before it is added up and settled through operator billing at the end of the month.

The direct exchange of a service or merchandise for money occurs with the pay-now option. An immediate exchange is created in the case of real cash payment, whereas the electronic pay-now payment by mobile phone implies a slight technical delay. Nonetheless, the compensation of values has a largely immediate character in comparison to other electronic payment methods, and occurs near real time. Payment by debit card is a pay-now option. However, in the case of a payment by EC card, actual monetary compensation only happens after approximately one working day due to the process involved.[130] In pay-now procedures, risk sharing between merchant and customer is balanced as none makes a risky advance service. If there is no credit balance on the debtor's account, card payment is interrupted immediately. The direct withdrawal of the amount from the bank account is

---

[130] Cf. Szameitat, 2001, p. 32

strongly preferred by 63.1% as a payment procedure for mobile macro-payment procedures.[131]

2.2.3.4 Place of Application

In the case of mobile payment applications for smartphones or tablet PCs, we also distinguish between server-based applications, which are activated through the mobile browser, and local applications, which are stored on the terminal.[132]

A server-based application requires access to mobile Internet in order to be used. Use in offline mode is not possible. These applications are also called web apps. Server-based mobile payment applications stimulate m-commerce via mobile Internet. Since 2012, Facebook has been working on the establishment of a browser-based mobile payment solution for users and developers.[133] For this purpose, Facebook initiated global cooperation with mobile providers for the development of a one-step mobile payment system. They are planning a SDK for developers, which enable a simple integration of the browser-based mobile payment solution.

The trustworthiness of a mobile payment system depends to a large extent on its data security. With mobile web apps, the security focus is on the

---

[131] Cf. Khodawandi/Pousttchi/Wiedemann, 2003, p. 12
[132] Cf. Bremmer (03 Nov. 2011): Native App oder mobile Website? (Native app or mobile website?)
[133] Cf. Lawson (27 Feb. 2012): Facebook Pushes for HTML 5 Standardization

transfer of the payment data and their storage on a server. With web applications that are based on HTML, CSS, and JavaScript, SSL certificates are frequently used for the encryption of the data transfer. Since the application is limited to a mobile web browser, it does not obtain any critical access to the local storage capacity of the device.[134]

In contrast to the server-based application, the local or native application is not activated through the browser. It is an independent program, which is stored on the device. A native application for the iOS operating system, for example, is based on Objective-C, and upon installation it can access iPhone sensors and applications. This includes access to the accelerometer, compass, contacts, camera, or GPS to locate the device. If the mobile payment app needs access to the camera function because the QR scanner has a central role in the payment process, it is recommended to implement the application as a native application. Current smartphones and tablet PCs offer sufficient memory for encrypted data storage. Apart from encrypted data storage, password protection is often chosen as an additional protection feature of a mobile payment app.

---

[134] Cf. Allan, 2012, p. 2

2.2.3.5 Applications in Commerce

Another factor in the distinction of mobile payment solutions is the concrete application scenario in commerce. With regards to the employment of mobile payment, Pousttchi sees the following possibilities: mobile commerce (MC), electronic commerce (EC), stationary merchant automat (SMA), stationary merchant person (SMP) and customer to customer (C2C).[135]

Mobile commerce (MC) stands for the acquisition of services and products via a mobile terminal. One of the first mobile products was, among other things, value-added services, such as mobile applications, games, news, or songs. Mobile providers offered payment methods like Premium SMS, WAP Billing or Direct Billing for their purchase. However, the reputation of Premium SMS and WAP Billing suffered because of high costs and non-transparent business models. Mistrust was created by cases in which clients signed subscriptions without knowing about it.[136] The so-called bill shock was a frequent result.[137] But the clients' trust was recovered by the transparent billing method of app store operators like Apple. The clients reacted with exponentially growing download figures in Apple's App Store.[138]

---

[135] Cf. Pousttchi, 2004, p. 56
[136] Cf. Chip Online (28 Dec. 2009): So bekommen Sie Ihr Geld zurück (This way you will get your money back)
[137] Cf. Meerman Scott, 2011, p. 128
[138] Cf. Tiefenthäler (05 March 2012): Apple: More than 25bn

Electronic commerce (EC) includes all types of digital business-to-consumer (B2C) commerce per stationary Internet. A solution for the expansion of online commerce by a mobile payment component is mpass, a joint project of Vodafone, o2 and Telekom.[139] The only condition to use mpass is to provide a German mobile phone number and a German bank account. Upon registration, the client obtains a personal PIN, which he uses in the online shop together with his mobile phone number. He then receives an SMS with a dynamically generated TAN, which has to be entered in the online shop to confirm the payment process. Finally, the amount is debited to the linked bank account through direct debit. This payment procedure has not been very successful yet, which is likely due to internal organizational issues.[140]

Stationary merchant automat (SMA) describes the classical payment transaction between a client and a machine. The machine acts as the merchant's representative. Application examples are the purchase of coffee, train tickets, or parking tickets. The amounts are often small, which means that mobile payment would mostly replace cash in this case. Due to the missing contact with a person at

---

downloads in App Store
[139] Cf. mpass
[140] Cf. Greif, (16 Aug. 2011): Telekom, O2 und Vodafone treiben Handybezahldienst Mpass voran (Telekom, O2, and Vodafone drive the mobile phone payment service Mpass)

the POS and an often time-critical buying situation, the process must be designed very clearly and transparently in order to gain the customer's trust and acceptance.

Stationary merchant person (SMP) describes the use of mobile payment in physical commerce. In this case a cashier manages the payment procedure at the POS. He guides the customer through the payment procedure. This person's qualified interpersonal interaction and benevolent presence indirectly has an immediate effect on the trustworthiness of the new payment procedure perceived by the client.[141]

Another application for mobile payment is the customer-to-customer (C2C) payment procedure – payment between two private individuals. In most cases, they either repay money they previously borrowed or they split a bill. This type of service is offered by Venmo, a start-up from New York City.[142] It focuses on the settlement of liabilities between friends. Venmo even has a "Trusted" status for friends, which – as soon as mutually confirmed – triggers an immediate money transfer between their Venmo accounts as soon as one of them requests money. Verification or approval by the debtor is dropped.

---

[141] On positive and negative spillover effects cf. Huber/Meyer/Vogel/Zimmermann, 2009, p. 31 f.
[142] Cf. Venmo

2.2.4 Mobile Payment Process

The mobile payment process is a complex process that involves different agents and responsibilities, depending on its technical construction. In order to better understand the mobile payment process, we will start with an explanation of the classical electronic payment process.

2.2.4.1 Agents

Five agents participate in the classical electronic payment process: card licensor, issuer, consumer, merchant, and acquirer.[143] (Cf. Abb. 14.)

---

[143] Cf. Deutsche Bundesbank: SEPA – Der einheitliche Eurozahlungsraum

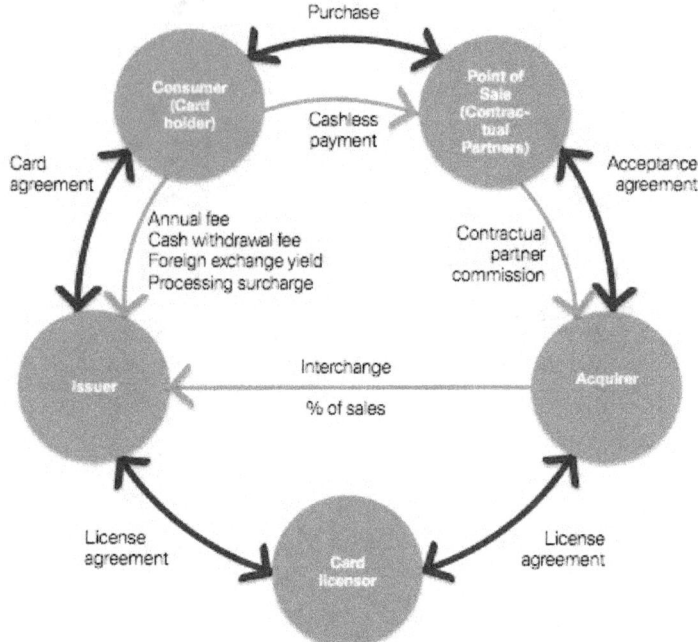

Fig. 14 – Agents in electronic payment transactions

Card licensors own the rights of the card brand, as in the case of Visa or MasterCard. They determine the procedure of the payment flow and are a direct contractual partner of the issuer and the acquirer.

Issuers are the issuers of the card, and they gave a license agreement with the card licensor. They offer the payment medium to the final consumers and at the same time guarantee its functionality and account management. The issuer can be a card organization, a bank, or a commercial enterprise like Amazon.

Consumers order the payment tool from issuers, and as a result both are in a contractual relationship. On the basis of the conditions agreed on, they receive a plastic card as a payment medium and can use it to settle liabilities with the merchant.

Merchants are the terminal or payment service provider's contractual partners and thus become the official point of acceptance of the new payment medium. They are also the acquirers' contractual partners.

Acquirers are the financial institutions that eventually claim the merchant's sales with the issuers' financial institutions and credit them to the merchants' accounts.

2.2.4.2 Value-Added Chain

In payment handling there are standardized procedures that are also valid for the mobile payment process. The classical electronic value-added chain has eight phases: registration, initiation, authentication, authorization, capturing, clearing, settlement, and administration.[144]

---

[144] In this work, the steps of the value-added chain cannot be described in detail, but only roughly. On value-added chain in mobile payment processes, cf. Contius/Martignoni, n.y., p. 61

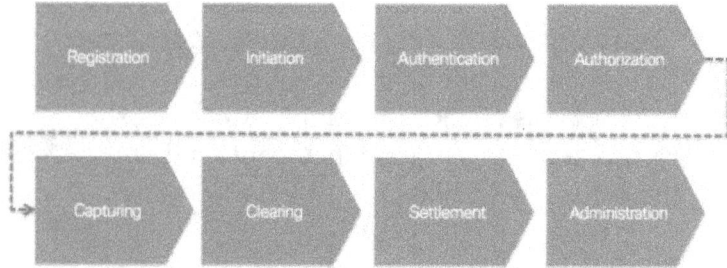

Fig. 15 – Mobile payment value-added chain – the eight key activities

Registration means the user signs up for the payment medium with the issuer or the mobile payment service provider. These can be banks or financial institutions. During this process step, a person is identified in order to guarantee authenticity and to lower the risk of fraud or money laundering. The quick achievement of critical mass at points of acceptance and users is a critical success factor for a new mobile payment service.

The initiation of the money transfer begins with a purchase situation. For this purpose, an electronic connection between client, merchant, and mobile payment provider is established in order to determine the contractual partners and to transmit the transaction data.

The authentication step verifies if the payment medium is valid and can be used for payment purposes. By entering his PIN or signature, the client confirms that he is the legal owner and gives his consent to the payment transfer.[145] A credit check is run In the case of a pre-paid payment process. Authentication is central to any payment process and is always carried out by the payment service provider. In mobile payment processes, the mobile phone number can be used as a new authentication feature.

Upon successful verification, the next step is the authorization of the transaction. At this moment, the payment is accepted by the payment provider and released by the system. With payment release, the payment provider grants the merchant a guarantee to pay. In the case of card abuse, he would assume the credit risk.

In the capturing phase, the actual payment order between the parties is generated.[146] With a credit card payment, for example, the merchant's bank (acquirer) makes a credit entry of the sales, and simultaneously he invoices the card issuer.

During clearing all relevant payment data is transferred between the banks. The payment data is exchanged in order to secure the claim of final payment settlement.[147]

During settlement the actual monetary compensation of the obligations between the parties

---

[145] Cf. Wolff, 2002, p. 64
[146] Cf. Wolff, 2002, p. 66
[147] On clearing and settlement processes cf. BIS (2000), p. 6 ff

occurs. The administrative processes and tasks with regards to the client, such as invoicing or client service, are handled within the framework of the administration of the payment process.

2.2.4.3 Remote and Proximity

The mobile payment process essentially resembles the traditional electronic payment process in its nature and processes. Nevertheless, the type of stakeholders and technical procedures change the structure of the service. Correspondingly, the payment process differs depending on whether it is proximity or a remote procedure.

The proximity payment process, for instance, requires two basic components: An NFC chip in the device, which stores the client's financial data, and a POS payment terminal that can read the NFC chip. This payment process then consists of the following steps:[148] (1) the client puts the NFC-compatible mobile phone within the readable range of the POS payment terminal. (2) The terminal recognizes the NFC chip, receives the device and transaction data and forwards them to the merchant's bank (acquirer). The merchant's bank transfers the transaction data to the client's bank (issuer), (3) which decrypts the data in order to identify the client's account and to authenticate the transaction. (4) In case of a positive authentication, the amount

---

[148] Cf. Shane, 2007, p. 218

is released to the acquirer per authorization order by the issuer.

Especially in the payment process with NFC-compatible devices, the circle of stakeholders had to be extended by device manufacturer, NFC chip manufacturer, mobile provider and mobile payment service provider (MPSP). The device manufacturer produces mobile phones and implements the technology supplier's NCF chip. The mobile provider sells the NCF-compatible devices to his final customers, while the mobile payment service provider (MPSP) develops mobile applications that enable the use of the mobile payment service.

The remote payment process distinguishes itself from the proximity payment process insofar as it can take place regardless of location via SMS or mobile Internet. An SMS-based payment process like PayPal's is divided into three steps[149]: (1) the client registers with PayPal and creates an account, which he links to a reference account. (2) The client then sends an SMS with the amount payable and the receiver's phone number to a short code (10.90 to 21225551981 to 729725). (3) Thereafter, PayPal sends a confirmation message to the sender and requests the personal identification number (PIN) for authentication. (4) Then, the client sends the PIN to the identical short code to release the payment. (5) The MPSP now immediately transmits

---

[149] Cf. PayPal: Texting with PayPal

the released amount to the receiver's PayPal account.

Although the steps in the value-added chain are maintained, the procedure is adapted depending on the technical components or stakeholders in the mobile payment process.

2.2.5 Stakeholders

According to Henkel, five groups determine the market situation of a mobile payment service: financial service providers (banks, credit card companies), mobile phone providers, mobile phone manufacturers, technology providers, and special mobile payment companies.[150] Lammer offers an extended and at the same time abstract perspective that includes final customers and merchants as stakeholders.[151] He defines final customers, merchants, and payment transaction providers as the three most important stakeholders and divides payment transaction providers into four subcategories: banks and credit card companies, mobile phone providers, mobile phone manufacturers, and technology manufacturers and specialized start-ups.

Based on Lammer, the stakeholders final customers, merchants, financial service providers, mobile phone providers, technology providers, and

---

[150] Cf. Henkel, 2002, p. 342
[151] Cf. Lammer, 2004, p. 130 ff

specialized mobile payment start-ups will be presented below. (Cf. fig. 16.) The perspective of the stakeholders will be described and explained with examples.

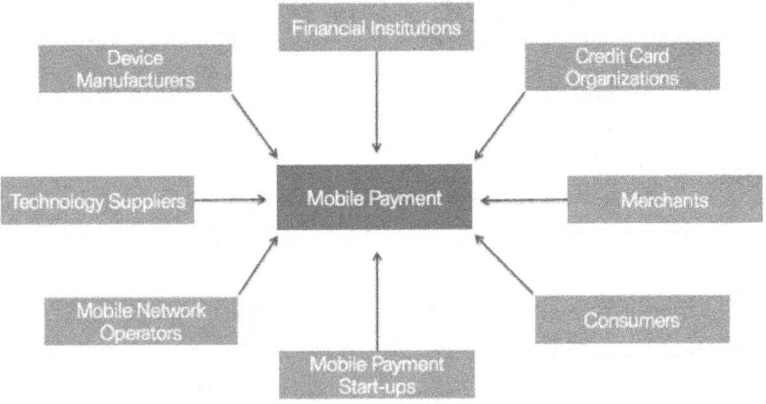

Fig. 16 – Mobile payment stakeholders

2.2.5.1 Customers

The final customers determine the success or failure of a mobile payment solution. Only through willingness to adopt the service and achieving a critical mass is the long-term existence of a service is guaranteed. This is why the final customers' interests and adoption criteria are so important. In the case of financial services, the trustworthiness of a service operator is a decisive adoption criterion for the final customer, as confirmed by a GfK study.[152] As a trust giver, the user assesses the

---

[152] Cf. GfK Study (12 May 2011): Mobile Payments: Trust and familiarity are crucial in driving adoption

credibility, security, and subjective meaningfulness of a service.

The GfK study found that 62% of the respondents rank mobile payment services as very useful.[153] In the groups of younger consumers aged from 16 to 24 (75%), innovators/early adopters (74%), and smartphone owners (72%), a significantly higher value was achieved.

Interestingly, perception varies in international comparisons. This is mainly due to the differences in payment infrastructure. Final customers in modern societies enjoy a considerably larger choice of payment options than consumers in developing countries. A U.S. citizen on average has 2.53 credit or debit cards, while a Chinese citizen only has 0.03 credit or debit cards.

Therefore, the willingness to accept or the basic general interest in mobile payment services is noticeably higher in countries like China (82%) and Brazil (73%) than in comparably more developed regions such as the USA or Europe (50% on average).[154]

---

[153] Cf. GfK Study (12 May 2011): Mobile Payments: Trust and familiarity are crucial in driving adoption

[154] Cf. GfK Studie (12.05.11): Mobile Payments: Trust and familiarity are crucial in driving adoption, key word "Global Preferences"

2.2.5.2 Merchants

However, before the final customer can be enthused for such a service, merchants must be attracted as intermediaries and points of acceptance. A high density of points of acceptance is a significant success factor for the expansion of a new mobile payment service. The merchant has the role of the first expert assessment of the service in trust generation with the final customer. The merchant thus acts as a trustworthy intermediary. His credibility and competence have a signal effect and a direct positive impact on the trustworthiness of the service.

The trustworthiness of a mobile payment service is also important for the merchant's reputation. Technical security and advantage must therefore be visible for both the merchant and the final customer. Before a merchant utilizes a new payment service, he will examine its strengths and weaknesses. Relevant advantages are lower costs per transaction, faster client checkout, higher client satisfaction, higher sales, lower bad debt losses, lower fraud rate, and cost savings through reduced cash trade. Additional success factors are simple implementation and the compatibility with other technical standards or interfaces.

2.2.5.3 Financial Service Providers

For financial service providers, mobile payment represents the next logical step in adding value to their service portfolio: mobile payment provides an additional channel within the provisions of their own systems of payment. Relevant financial service providers include banks and credit card issuers.

For banks, the processing of payments is part of their core business, comprising over 30% of total turnover.[155] Mobile payment could therefore be regarded as an addition to their existing range of products. Having their own mobile payment service would therefore serve to tighten customer ties with banks and would enable the development of new sales channels. In addition, the possession of its own mobile payment service at an early stage in the market could serve to improve a bank's public image or position it as an innovator. But, even though 74.7% of final customers regard banks as the most trustworthy providers of mobile payment services, banks themselves as the most reluctant participants in the market when it comes to the provision of new mobile payment solutions.[156]

Credit card issuers, such as Visa and MasterCard, by contrast, are drivers of the mobile payment market. For them, mobile payment is not simply a feature

---

[155] Cf. Boston Consulting Group, 2012, n. p.
[156] Cf. Wiedemann/Goeke/Pousttchi (o. J.): Ausgestalltung mobiler Bezahlverfahren

but a product in itself. Mobile phones have become substitutes for plastic cards. Credit card issuers have been involved in pilot programs with mobile phone and equipment manufacturers for years. Visa has already established numerous pilot projects with banks and mobile phone providers worldwide.[157] In February 2012, Visa even announced a worldwide strategic joint project with Vodafone to develop a joint mobile payment solution.[158] The use of this NFC technology at POS would be an extremely promising development.

2.2.5.4 Mobile Phone Providers

With their broad client bases and their extensive experience in the area of telecommunications, mobile phone providers seem to have been designed to provide mobile payment solutions. As network operators who have to calculate charges for speech and data services, they have experience billing clients using their own billing systems. They are particularly familiar with trans-network and cross-border clearing and billing processes.[159]

Mobile phone providers regard mobile payment as an opportunity to diversify their business model and reach new areas of operation by, for example, combining electronic and stationary trading.[160] In

---

[157] Cf. Visa: Visa Mobile Payment Pilots. Mobile Fact Sheet Chart
[158] Cf. Mansfield (27.02.12): Vodafone Teams Up with Visa for Global Mobile Payments Push
[159] Cf. Taga/Karlsson 2004, p. 6
[160] Cf. Arthur D. Little, 2009, p. 12

addition to attracting new clients, they need to strengthen ties with existing clients by extending the range of their services. As infrastructure operators with a broad client base, they would particularly benefit from the mobile commerce joint turnover model.[161] Charging for mobile purchases using the mobile service provider's own billing service has a clear advantage: the customer doesn't have to register for these purchases separately, since he or she can be identified by an MSISDN (Mobile Station Integrated Services Digital Network Number). This makes it unnecessary to fill out individual payment details for each purchase.

Mobile phone providers in Germany have an ARPU (Average Revenue per User) of approximately 20 EUR and are therefore primarily accustomed to processing micropayments.[162] As a result, they have limited experience in handling macro payment processes. Macro payment processing involves a higher credit and fraud risk and therefore requires bank-like risk management strategies. In addition, mobile phone providers usually only have experience with charging, not with the actual transfer of money from a customer's to a merchant's account.

---

[161] Cf. Costello, 2002, p. 12
[162] The mobile phone providers in Germany by ARPU: T-Mobile (€22.90), Vodafone (€21.90), E-Plus (€17.20), o2 (€19.40) See The Netsize Guide 2011: Truly Mobile, Gemalto, p. 126

In order to handle actual payments, mobile phone providers need their own banking license as specified in § 32 of the KWG (German Banking Act). This involves strict stipulations and requirements, such as personal eligibility and a minimum capital endowment. Most mobile phone providers, despite the strategic benefits of mobile payment, do not have a banking license, since the establishment and maintenance of a bank-like department is a costly and time-consuming process. Joint projects and takeovers are frequent, as they are a more effective means of achieving the same ends. For example, in late 2011, the telecom company Net Mobile purchased the bank Werther AG, listed on the stock exchange, together with its full banking license.[163] Strategic purchases of this kind enable mobile phone providers to make preparations for the implementation of their own mobile payment systems with billing handled in house.

2.2.5.5 Technology Providers

For the purposes of this thesis, technology providers are defined as companies that contribute to mobile payment solutions by manufacturing necessary technical components. They have no wish to take a central role in the establishment and development of the ecosystem.[164] However, they influence the success or failure of mobile payment. Mobile payment systems will only be able to reach a

---

[163] Cf. Sommer (21.12.11): Telefonkonzern kauft sich eigene Bank
[164] Cf. Henkel, 2002, p. 345

substantial market quickly if the necessary technology is developed and distributed. Technology providers include mobile phone, software, SIM card, chip and terminal manufacturers.

Mobile phone and software manufacturers, in particular, regard mobile payment solutions as an additional service that will increase customers' use of their products. Mobile payment also makes it possible to extend the use of their products to offline interactions. This will provide mobile phone and software manufacturers with more detailed customer data, which could be of use in their own product development or in selling their own media to advertisers with certain selection criteria.

In 2009, Nokia established the service Nokia Money. This service was aimed primarily at people without bank accounts, who were mostly to be found in developing countries.[165] In order to implement Nokia Money's prepaid payment model, they entered into a strategic partnership with obopay. The idea was that users should upload money to their mobile phone accounts in advance; in order to transfer money later using encrypted text messages. Recipients were identified by typing in their mobile phone numbers.

---

[165] Cf. Müller (03.09.09): Nokia money: Bezahlen mit dem Prepaid-Handy statt einem Konto

In 2010, software and hardware manufacturer Apple secured a patent for an NFC payment technology. The patent describes the process of transferring credit card information between a device and an iTunes account.[166] Experts currently predict that the iPhone 5 will include an Apple-specific mobile payment solution.[167]

The software manufacturer Google also founded its own mobile payment solution in 2011, under the name Google Wallet.[168] It allows customers to use their android-based smartphone and wireless NFC communication technology to pay in shops. The fast food chain Subway, the department store Macy's and the clothing store American Eagle Outfitters are participating in the pilot program.

The manufacturers of payment terminals have an important role to play in the implementation of NFC technology to make mobile payments at POS. For example, MasterCard PayPass terminals were used in the Google Wallet pilot project. There are a large number of POS terminal manufacturers, such as the Chinese terminal manufacturer Hangzhou Maizhixibo. The company has been researching RFID technology since 2005 and currently offers

---

[166] Cf. Free Patents Online (23.08.11): United States Patent Application US20110269423

[167] Cf. Graziano (31.01.12): Apple taking NFC payments mainstream with iPhone 5

[168] Cf. Sawall (25.05.11): Googles Handy-Bezahlsystem wird morgen vorgestellt

NFC-capable terminals. EKEMP Electronics Ltd. also develops and sells POS terminals that incorporate RFID technology, permitting mobile payments.

In addition to terminal manufacturers, SIM card and chip manufacturers are further examples of technology providers. SIM cards, the most important means of identification for mobile phone customers, already play a central role in the mobile communication process. They are now also being considered as possible secure elements or channels of RFID technology. The German supplier Giesecke & Devrient has already developed NFC-capable SIM cards, which are currently being used in a mobile payment pilot project with MVNO Nòverca in Italy.[169] SIM card manufacturers work closely with equipment manufacturers and mobile phone providers, since, as their major buyers, the manufacturers and providers ultimately determine what features the smartphones will have.

Chip manufacturers are also stakeholders because they equip devices with the necessary technology and significantly determine the performance of the devices. Chip manufacturer Texas Instruments introduced the new WiLink 8 Chip for mobile terminals at the Mobile World Congress in 2012.[170]

---

[169] Cf. Schenk (23.02.12): G&D präsentiert NFC-fähige SIM-Karten
[170] Cf. Donovan (27.02.12): Texas Instruments Announces New Partnerships For OMAP 5

This chip has the potential to speed up the market demand for NFC technology in mobile devices since, together with many other transmission technologies; it delivers an NFC component that is compact and conveniently standardized.

2.2.5.6 Specialized Mobile Payment Start-ups

In addition to the classic players in the mobile and finance branches, there are now specialized mobile payment start-ups in the mobile payment marketplace. Their central business idea is based on the development and marketing of mobile payment solutions. Financed by venture capital firms, they are often highly innovative and are also at high risk of failure. Their limited budgets, fixed for each round of financing, oblige them to act quickly and efficiently. This frequently leads to radical innovations, which can be quickly implemented and creatively developed.

Square, a start-up, announced its first mobile payment solution in 2010. This enabled any user to accept credit card payments using a smartphone app, by either keying in the credit card details manually or by using a card reader dongle to obtain them. The Square card reader dongle is concealed in the phone's audio exit. The customer confirms the payment by signing on the touch screen of the mobile terminal with his finger. The target group for this service are, in particular, small and very small businesses, for whom the purchase of a classic credit card reader would seem uneconomical or impractical, because they run a mobile business.

Another mobile payment start-up, Venmo, focuses on money transfers between friends. A mobile app allows users to transfer money to friends. The money transfer always takes place via the user's own Venmo account. To use this service requires a US bank account and a US mobile phone number. The user can then pay the amount owed using a credit card, bank account or Venmo prepaid account.

It's particularly challenging for new, young technology companies to gain users' trust. Square has already had to conduct several public debates on security with its competitor VeriFon.[171] For that very reason, particular emphasis is placed on security at Square and the topic is often discussed.[172]

Venmo has approached the topic of trust in a very proactive manner; by introducing a relationship option it calls trust.[173] Both users have to accept this option, which then reduces the number of security steps involved in a transaction. If a trustworthy friend requests a sum of money, that money will immediately be deducted from the donor's account and credited to the recipient's.

---

[171] Cf. Olivarez-Giles (10.03.11): Square answers VeriFone's accusations on security of mobile credit card reader
[172] Cf. Square: Security
[173] Cf. Venmo: How and why to use "Trust"

2.2.6 Summary

Global distribution has turned the mobile phone into the seventh mass medium. But the mobile phone is different from all previous mass media because of the intimate significance it has for its owner. There is no item of mass technology we feel closer to than our mobile phones. The extent of our identification with our phones has increased even further because of the numerous ways of personalizing the device. Further developments in design, usability and technological safety have also made it into a serious vehicle for mobile payment solutions. As with all other payment channels, when it comes to mobile payment services, trust is a prime requirement before users will make active use of the service.

Investigating mobile payment reveals it to be a highly complex ecosystem. Numerous stakeholders, extensive classification options and complex payment processes have resulted in numerous mobile payment solutions in the marketplace, leading to uncertainty and lack of transparency for users. The role which trust plays in a mobile payment service and which aspects of trust need to be taken into account in its development and implementation largely depends on the risks the users have to sustain. For mobile payment services, these are risks like fraud, theft or data abuse. Market research has revealed two fundamental consumer anxieties: how can the payment details be protected in the case of the loss or theft of the device and how can they be reactivated and how secure is the transaction process in which payment details are transferred between the sender and the recipient? User anxieties about the trustworthiness of the system arise particularly frequently in respect to data transfer, data storage, service availability and transparency. End users of mobile payment services list factors such as reputation, public profile, safety standards and the number of points of acceptance as critical indicators of trustworthiness.

Part 3 will offer a practical comparison, in which representatives of current mobile payment services will answer questions about how they take account of the trust factor in both the way their systems are constructed and in how they communicate with the public, which aspects of trust are particularly important and what measures they take to fulfil them.

3 Practical Comparisons

3.1 Introduction

Having examined these theoretical considerations, this thesis will now go on to conduct a practical comparison. Company representatives were interviewed about the role of trust in the development of mobile payment services. The companies involved were, at the time, developing or employing state-of-the-art solutions in the mobile payment marketplace.

Anyone who has observed the current market activity in the mobile payment sector will have noticed how dynamic the sector is. Numerous mobile payment solutions are pouring in to the market. And its not just companies from the traditional field of finance that are competing for business – companies from the unrelated field of technology are also developing innovative solutions. Despite the variety of different mobile payment solutions on the market, the table below (Fig. 17) is an attempt to sort them and provide a rough overview of the market.[174]

Category	Mobile Payment Solutions
Card Reader Solutions	Square, PayPal here, goPayment, payware mobile, Zenpay, SumUp, SalesVu, payanywhere, AppCharge.

---

[174] This representation makes no claims to completeness. The intention is only to give an impression of the many technical solutions, which were on the market at the time of writing.

Mobile Wallet	ISIS (NFC), Google Wallet (NFC), Microsoft Wallet (Mobile Internet), Apple Passbook (mobile internet), Crowdmob, LevelUp, Vemo (mobile Internet), PayPal (card.io), mWallet (NFC), digimo (NFC), cashlog, Sprint Touch (NFC), clover (mobile internet), kuapay, dwolla, Wallby.
In-App Solutions	Starbucks, Walmart, Burger King, Touch&Travel, AirPlus.
QR Code	PayPal QR Shopping, Mr Net Group, Cleap, digimo.

Fig. 17 – Mobile payment solutions arranged by technological category.

The increasing complexity of the market seems likely to lead to increased anxiety among customers, making the trust factor even more crucial.

The following practical comparison will focus on three companies. The companies chosen were all important driving forces in the implementation of mobile payment solutions in the German market at the time when this paper was written: Deutsche Telekom, Vodafone D2 and PayPal.

The interviews were conducted with company representatives from the German market. The questions took the form of qualitative, semi-open interviews. The aim of the interviews was to find out how the company representatives viewed and how important they found the question of trust in the design of their mobile payment services and in their communications with the public.

3.2 Telekom

With 93 million mobile phone customers in Europe and the US, Deutsche Telekom AG is one of the leading service companies in the telecommunications and information technology sectors.[175] Telekom is an offshoot of the state-owned company Deutsche Bundespost. In 2012, the company continued to place transformation at the heart of its business strategy.[176] The telecommunications market demands a very high degree of flexibility and innovation from companies. Transformation in the sense of change implies a strategic appeal for continual rethinking and the transition from current to future business models.

3.2.1 Deutsche Telekom's Payment Ecosystem

Deutsche Telekom clearly treats mobile payment as a strategic business field. This is clear not just from their website, but also from their strategic partnerships, such as their partnership with the

---

[175] Cf. Deutsche Telekom: Annual Report 2011
[176] Cf. Deutsche Telekom: Business Presentation 2012

MasterCard corporation, begun in July 2012.[177] A study commissioned by Deutsche Telekom also states mobile payment as an upcoming strategic business area. 66% of the Germans and 86% of the Americans, who took part in the survey, were convinced that mobile payment would become widely accepted.[178] The study also reached some interesting conclusions as to the kinds of goods that customers wish to purchase by mobile payment. Together with minor purchases and everyday goods, a high percentage of survey participants also mentioned clothes, tickets, gasoline and durable consumer goods. This demonstrates a wide range of uses for mobile payment.

This is perhaps one reason why Deutsche Telekom has decided on the implementation of a ‚mobile wallet', instead of a single payment solution.[179] The mobile wallet (also called m-wallet) solution is, however, only one component of Deutsche Telekom's payment ecosystem. This will also include "Online Payment" through the subsidiary ClickandBuy, payment cards issued by their strategic partner MasterCard und POS payment terminals with NFC readers available for transactions. In these ways, Deutsche Telekom plans to take an active part in the further development of the

---

[177] Cf. Deutsche Telekom: Mobiles Bezahlen; Deutsche Telekom (02.07.12): Mastercard und Deutsche Telekom
[178] Cf. Deutsche Telekom (02.07.12): Studie Mobile Payment
[179] Cf. Deutsche Telekom (02.07.12): Telekom stellt sich für den Payment-Markt auf.

payment market and provide simple and secure payment solutions for mobile, online and POS transactions.

3.2.2 The Mobile Wallet

At the time of the interview, Telekom Innovation Laboratories, T Labs for short, were developing a mobile wallet (m-wallet) solution.[180] The interview took place with the project manager responsible for the m-wallet, Zhiyun Ren. He provided insights into the service's current state of development and into the importance of the trust factor in T Labs' mobile payment solution.

Mobile wallet solutions are not simply about digitizing a method of payment. They constitute a multi-application platform. The platform is installed with various applications such as payment, ticketing, coupons, loyalty and access keys (e.g. car or house keys).[181] Different kinds of cards can be activated for this purpose, such as credit, debit, prepaid, customer loyalty cards, entrance tickets or train tickets. The payment functions of the m-wallet include both remote (via mobile internet) and proximity payments using NFC technology. Universal integrated circuit cards (UICCs), the new generation of SIM cards, provide the secure

---

[180] Cf. Telekom Innovation Laboratories: Mobile Wallet
[181] Cf. Ren, Zhiyun (2012). Personal interview, conducted by the author, Berlin, 8 June 2012, 0:20 min.

element.[182] This new kind of SIM card offers a number of new safety features for data storage and encoding on the card's chip. This enables secure payment and identity management functions to be carried out on the mobile phone.[183]

At the time of the interview, the m-wallet solution was in the pilot phase. The first product launch is to take place in Poland in 2012.[184] A product launch in Germany is planned for some time in the course of 2013.

The m-wallet will ultimately be a platform to which service providers will be able to attach the services they offer, through standardized interfaces. The standardizing of the interfaces began with the ‚E5' association, a group of Europe's largest mobile phone service providers, made up of Deutsche Telekom, Vodafone, Telekom Italia, Telefonica and Orange. Because of concerns about cartel-related legal issues, the E5 were asked to hold the discussion about interfaces jointly with the GSMA, so that smaller mobile phone service providers could also be involved in the discussion.[185]

---

[182] Cf. Ren, Zhiyun (2012). Personal interview, conducted by the author, Berlin, 8 June 2012, 2:50; 5:16 min.
[183] Cf. Telekom Innovation Laboratories: Mobile Wallet - Die digitalisierte Brieftasche
[184] Cf. Ren, Zhiyun (2012). Personal interview, conducted by the author, Berlin, 8 June 2012, 6:57 min.
[185] Cf. Computerwoche (14.03.12): Europas Telekomriesen unter Verdacht

3.2.3 An User-Oriented Approach

According to Ren, trust is regarded as an important factor in the implementation and distribution of a mobile payment service.[186] Many people have little experience with the application or use of NFC technology and can quickly become unsettled by the speed of transactions. Telekom hopes to reassure customers with a user-oriented approach and win their trust through control and transparency.

The unified user interface is designed to give the user a feeling of security, as it uses the familiar analogy of the stock market and incorporates and unifies different payment and authentication services. The design of the service – in the sense of the user interfaces, interactions, processes and informational architecture – will be tested in a three-stage model before its launch. First, user interface experts will evaluate the wireframes. Then mock-ups will be constructed. Early adopters will test these and then, after reworking the model, the first programmed prototype of the app will be evaluated by average users.[187]

The user-oriented approach also aims to allow the customer to personalize the application's settings. This should have a substantial effect on encouraging customer confidence. The customer can decide

---

[186] Cf. Ren, Zhiyun (2012). Personal interview, conducted by the author, Berlin, 8 June 2012, 11:44 min.

[187] Cf. Ren, Zhiyun (2012). Personal interview, conducted by the author, Berlin, 8 June 2012, 19:25 min.

which payment methods or which level of security should be set as the default. If you choose the fast pay method, the mobile phone will only need to be placed on the payment terminal once in order to pay the bill, with no further confirmation required. But you will also be able to incorporate an automatic PIN number request and require authentication for each payment.

Clearly, this requires a self-explanatory user interface. If there are numerous different settings, a confusing user guide could lead to distrust and therefore to a rejection of the service. This also raises the question of how much control the user actually wants to have over the device and at what point the administrative effort required is too great and cancels out the convenience factor. This phenomenon can also be observed in the case of the social media platform Facebook. The high number and lack of transparency of the optional privacy settings can lead to suspicion and even to reduced use or to no further use at all.[188] A non-standardized design coupled with a high number of settings that are difficult to locate can turn the user's control over the device into an annoying and tiresome administrative task.

---

[188] Cf. Computer Bild.de (16.09.11): Neue Facebook-Einstellung: Freund, Fan oder Abonnent?

When it comes to building trust, control must always be accompanied by transparency.[189] A multi-application platform offers a multitude of possibilities and settings such as different payment methods, payment settings, safety levels, service programs, and password and card settings. The payment process, in particular, requires being able to keep clear track of present and past transactions, both at the moment of payment and afterwards. In addition to sensible and clear system-related feedback, the m-wallet application also keeps track of all transactions to date in a personal payment history. Transparency in the sense of traceability (being able to keep track) also requires additional services such as a display of the current balance or credit so that the user always knows his current financial position.

3.2.4 The Security Aspect

"Security is the top priority of our mobile wallet application," according to Ren.[190]

The prime importance of the security aspect in building trust is closely connected to the potential user risks associated with mobile payment. These include the real possibility that the user might lose

---

[189] Cf. Ren, Zhiyun (2012). Personal interview, conducted by the author, Berlin, 8 June 2012, 19:14 min.
[190] Cf. Ren, Zhiyun (2012). Personal interview, conducted by the author, Berlin, 8 June 2012, 6:38 min

his mobile phone as well as the fear of data theft or abuse.[191]

This anxiety is the result of the new ways in which mobile phones can be used and the increased personal importance of the phones. Phones, which were once simply used to make telephone calls, are now multi-functional personal devices. A great deal of personal information is now stored on smartphones. This makes the loss of a phone significant and even dangerous. Payment data are among the most sensitive of personal data. The fear that data could be stolen or abused is, however, often the result of an incomplete understanding of the technical possibilities or of a lack of familiarity with new technologies such as NFC.
Ren acknowledges that time is a vital element in the building of trust.[192] In particular, he talks about time in relation to the development of learning effects, experiences and habits with a new solution. This is, according to him, a long process of imitation, experience and evaluation. Perhaps time could be described as synonymous with patience in order to see things from the user's perspective. Users expect to have to have patience in the sense of undergoing a personal learning process, in order to try out the new application and to come to understand it. It

---

[191] Cf. Ren, Zhiyun (2012). Personal interview, conducted by the author, Berlin, 8 June 2012, 23:42 min
[192] Cf. Ren, Zhiyun (2012). Personal interview, conducted by the author, Berlin, 8 June 2012, 32:39 min

takes time to learn the process and to find the optimum solution.

Telekom hopes to reassure its customers by including up-to-date data encrypting options on the device itself and on the SIM card.[193] For example, when you access the program, it requests your PIN number. Should the device actually get lost, you can block the personal m-wallet account and the associated payment cards by telephone and reopen it later using a backup service.[194]

Mobile payment also has to compete with traditional means of payment such as cash and plastic cards. The mobile payment process therefore needs to be at least as fast as these long-established alternatives. In the context of the normal duration of a payment process, speed can be seen as a hygiene factor. At the moment of payment, speed only has an influence on trust if there are delays in the payment process or if the duration exceeds or falls short of the norm. People notice the speed, if the process is too fast or too slow. When their attention is drawn to the speed, this often causes confusion or anxiety.

The challenge for the design of the service is to strike the ideal balance between security and speed.

---

[193] Cf. Ren, Zhiyun (2012). Personal interview, conducted by the author, Berlin, 8 June 2012, 23:42 min
[194] Cf. Ren, Zhiyun (2012). Personal interview, conducted by the author, Berlin, 8 June 2012, 30:04 min

Too much security will have a negative impact on the speed of the process and vice versa.[195]

3.2.5 Trust Threshold

According to Ren, the important factors in user trust are safety, time and brand recognition.[196] It is clear from the interview that establishing trust in technical systems is intimately connected with familiarity and learning processes. It would only make sense to introduce an NFC-based payment system once a certain level of market maturity had been achieved, in regard to the spread of technology and points of acceptance.[197] That's why they are going to launch m-wallet in Poland first. There are already approx. 80,000 points of acceptance in the Polish retail sector, as opposed to only around 10,000 in Germany.[198]

Points of acceptance and partners also play a role in increasing market confidence. Partners have the task of publicizing the new payment options and explaining them at the point of acceptance. The partners' reputations can have an indirect effect on the reputation of the mobile payment service itself. Publicizing the new payment options can have the

---

[195] Cf. Ren, Zhiyun (2012). Personal interview, conducted by the author, Berlin, 8 June 2012, 1:00:24 min
[196] Cf. Ren, Zhiyun (2012). Personal interview, conducted by the author, Berlin, 8 June 2012, 32:39 min; 41:31 min
[197] Cf. Ren, Zhiyun (2012). Personal interview, conducted by the author, Berlin, 8 June 2012, 41:40 min
[198] Cf. Ren, Zhiyun (2012). Personal interview, conducted by the author, Berlin, 8 June 2012, 44:01 min

effect of recommending them. The relationship of trust between the partner and the customer can be transferred at that moment to the mobile payment service. The partner therefore also acts as a social filter and enables the mobile payment service to gain visibility among the partner's customers.

There are, however, observable differences in the speed with which trust can be established.[199] First of all, there are differences between different regions. Each region has its own climate of openness or reluctance towards technology, depending on the cultural milieu. At the same time, there are also differences among different social groups and income classes. There are probably also differences in the willingness and speed with which people adopt new technology among different generations or age groups. This theory has received some recent support from the expression digital natives, coined by Marc Prensky.[200] The digital natives are defined as the first generation, which grew up with digital media such as the Internet, mp3 players, email, computer games and mobile phones as a matter of course. The expression "digital immigrants" on the other hand describes people who were born before 1970 and only learned to use digital media as adults. There is a big difference in the way these two groups use digital media or IT solutions. Their

---

[199] Cf. Ren, Zhiyun (2012). Personal interview, conducted by the author, Berlin, 8 June 2012, 53:55 min
[200] Cf. Prensky (2001): Digital Natives, Digital Immigrants

differences in technological knowledge and/or openness to technology lead to different levels of ease of acceptance and trust with regard to technical innovations.

3.3 Vodafone

The Vodafone group is one of the largest mobile phone service providers in the world, with 370.9 million customers.[201] Vodafone D2 GmbH originated as Mannesmann Mobilfunk. The Vodafone group took over the mobile phone branch in 2000.

3.3.1 The Company

With 36 461,000 mobile phone customers in Germany, Vodafone is the country's biggest mobile phone provider, followed by T-Mobile with 35 100,000 mobile phone customers (as of 31.03.12).[202]

In the most recent Vodafone group presentation, Vodafone group's CEO, Vittorio Colao, describes mobile payment as a strategic field of business.[203] The 2011 annual report also mentions mobile payment as a new service area in the context of the strategic thesis "Focus of key areas of growth potential (Mobile data, emerging markets,

---

[201] Cf. Vodafone Group Plc. (31.03.12): Annual Report 2011
[202] Cf. Bundesnetzagentur (2012): Teilnehmerentwicklung im Mobilfunk nach Netzen pro Quartal
[203] Cf. Colao (20.09.11): Vodafone Group Presentation. Bringing "Supermobile" to life

enterprise, total communications and new services)".[204] In that context, mobile payment is described as a mobile wallet solution that uses NFC technology.

3.3.2 Evolution of Mobile Payment

Marc-André Fengler, head of payments with Vodafone D2 GmbH has described the three evolutionary stages of mobile payment at Vodafone in an interview.[205] Like other mobile phone providers, Vodafone has tradition of involvement in mobile payment.

Mobile payment began at Vodafone in 2002 with payment 1.0, also known as mobile billing.[206] Mobile billing allows customers to pay either through their telephone bill or by using a prepaid voucher. At first, it was used primarily for the purchase of digital goods such as logos, ringtones or games. They later added mobile phone postal services, mobile phone parking and mobile payment for publications, such as articles. In 2008, they introduced payment 2.0 in the form of a mobile pass as for mobile extended payments for e-commerce.[207] The new mobile payment solution, the mobile wallet application that

---

[204] Cf. Vodafone Group Plc. (31.03.12): Annual Report 2011, p. 24
[205] Cf. Fengler, Marc-André (2012). Personal interview, conducted by the author, Düsseldorf, 21 June 2012, 1:12 min.
[206] Cf. Vodafone Deutschland: Mobil bezahlen
[207] Cf. Vodafone Deutschland: mpass – die sichere Bezahlmethode per Handy

uses NFC technology, will be the next stage of development, payment 3.0.[208]

They are currently implementing the mobile wallet application jointly with Visa. The beginning of their partnership was announced in February 2012 at the Mobile World conference in Barcelona. It will be the largest joint project in the field of mobile payment in the world.[209] The goal is to provide a mobile alternative to cash and cards. The mobile payment solution will be released as a Vodafone product and offered to Vodafone customers in more than 30 countries.[210] However, it will be underpinned by Visa's payment infrastructure.

Like Telekom, Vodafone's solution will also provide an open platform for mobile applications and will later be able to function as a mobile wallet on the smartphone. In addition to payment functions, there will be other applications, such as loyalty programs or coupons. The platform is to be open to partners from all branches of industry, such as finance, commerce and transport.

They have already completed the planning as to the solution at POS and are currently at the development stage with regard to the technical

---

[208] Cf. Vodafone Group Plc. (31.03.12): Annual Report 2011
[209] Cf. Vodafone press communication (27.02.12): Vodafone and Visa Announce World's Largest Mobile Payments Partnership
[210] Cf. Visa Press communication (05.03.12): Vodafone und Visa gehen weltgrößte Partnerschaft beim mobilen Bezahlen ein

solution. The goal is to enter the market as soon as possible.[211] They have not yet conducted any user tests of the product and therefore cannot evaluate the mobile wallet solution experience using that kind of data. They are evaluating the experience based on current mobile payment practices and market studies conducted by Vodafone.

3.3.3 Systemic and Perceived Security

Security plays an important role in the building of trust in the solution for Vodafone, just as it does for the other companies described here. Fengler even differentiates between systemic and perceived security.[212] Systemic security is connected with the concrete task of designing and constructing an ideal technical system, whereas perceived security can be achieved through communication and explanation.

Systemic security involves the area of data storage on the one hand and data transfer on the other. The Vodafone solution will involve storing the payment data on a new generation of SIM cards. These SIM cards incorporate a special data container to which even Vodafone does not have access. An independent third party, acting as a trusted service manager, will be the only one capable of encrypting and decrypting the data.

---

[211] Cf. Fengler, Marc-André (2012). Personal interview, conducted by the author, Düsseldorf, 21 June 2012, 6:33 min
[212] Cf. Fengler, Marc-André (2012). Personal interview, conducted by the author, Düsseldorf, 21 June 2012, 12:30 min

Another fear is that customer data could be read using NFC mobile technology on an RFID chip. There was a discussion about data protection in this context after the Sparkasse introduced a new, plastic card equipped with NFC.[213] The RFID chip continually transmits a unique identification number in a radius of 10cm – in accordance with ISO standard 14443 -- there is concern among data protection specialists that this might lead to the creation of anonymous movement profiles. Their fear is that hackers with homemade NFC readers would be able to withdraw sums of up to 20€ "in passing". However, a radius of 10 cm is very small and the technical outlay involved in reading the chips is so great that Vodafone does not see this as a risk. There are also no legal data protection issues or concerns, since data is always stored in encrypted form and it is not possible to trace it to specified individuals. Official certification of the process is crucial, according to Fengler, in order to guarantee the process's security.[214]

In addition to systemic security Fengler also emphasizes perceived security, which is primarily a question of achieving a balance between knowledge and non-knowledge. Ignorance and distrust are often the results of incomprehension. Increasing

---

[213] Cf. Reißmann (19.05.12): Unsichtbares Kleingeld verrät seinen Besitzer
[214] Cf. Fengler, Marc-André (2012). Personal interview, conducted by the author, Düsseldorf, 21 June 2012, 17:14 min

perceived security is therefore primarily a question of marketing and communication.

The challenge lies primarily in communicating sufficient information without unsettling users.[215] If you go into too much technical detail, you will lose your audience. The goal, therefore, is to give the customer a feeling of security by describing the technical mechanisms in terms that can be generally understood, without descending to the level of fine detail. Bear in mind, that if one communicates in terms that are too emotional or superficial, it can damage the credibility of one's information.

3.3.4 On Establishing of Trust

In addition to security, Fengler identifies further factors that can increase the trustworthiness of a mobile payment service. These include adhering to standards, reputation, integrity, usability and viral effects as risk-bridging strategies.[216]

A standard is a recognized way of doing something that has become established, such as keying a PIN number into an ATM when someone withdraws money. When designing a payment process, it can help establish trust if one incorporates learned sequences of actions and elements of established

---

[215] Cf. Fengler, Marc-André (2012). Personal interview, conducted by the author, Düsseldorf, 21 June 2012, 17:16 min
[216] Cf. Fengler, Marc-André (2012). Personal interview, conducted by the author, Düsseldorf, 21 June 2012, 25:01 min

payment processes to encourage people to accept the service. From the perspective of the merchant, standards can, for example, be defined in terms of technical interfaces. For the merchant, technical standards mean manageable outlays and technical security. Their primary goals are, namely, increased efficiency and cost savings. Therefore the aim is to incorporate standardized technical interfaces into the new technology to minimize outlay on integration.

A company's reputation, the name it has established for itself over time, is a form of social currency. Bourdieu defines reputation as a form of symbolic capital.[217] It is a form of social recognition, comparable with a kind of social advance payment issued against the guarantee of future popularity. The user of a new payment service will care about the service provider's reputation and the reputations of its partners. The partnership of Vodafone and Visa, in particular, involves two strong and long-established major companies working together. Both companies have a long history in the marketplace and are considered to be major players in their respective branches. The partnership unites telecommunications with the financial sector. But there are probably two factors involved in the evaluation of a new payment service: its general reputation in the market and the personal social endorsement of individual observers. Both of these

---

[217] Cf. Fröhlich/Rehbein, 2009, p. 138

factors can affect the big picture and strengthen or weaken the company's perceived reputation.

In this context, the integrity of the service and/or its providers is a very important part of its trustworthiness.[218] The administration of payment processes gives rise to especially sensitive data that need to be moved and processed. The final customer is likely to feel some anxiety when it comes to the responsible handling of his data. The integrity factor affects, in particular, the intactness of the data as well as the correct storage and handling of the data. Integrity is therefore intimately connected with safety. However, integrity has a strongly systemic, technical frame of reference and can scarcely be experienced as an emotion.

User-friendliness or usability also contributes to a service's trustworthiness.[219] According to the official norm DIN EN ISO 9241 usability should increase effectiveness, efficiency and satisfaction.[220] In this context, Fengler emphasizes the fun factor in trust building.[221] It's crucial that the user should experience the service as something that facilitates daily life and that he or she should enjoy using it. The initial moment of use determines whether the

---

[218] Cf. Fengler, Marc-André (2012). Personal interview, conducted by the author, Düsseldorf, 21 June 2012, 23:40 min
[219] Cf. Fengler, Marc-André (2012). Personal interview, conducted by the author, Düsseldorf, 21 June 2012, 27:18 min
[220] Cf. ISO Standard. Search #: 9241
[221] Cf. Fengler, Marc-André (2012). Personal interview, conducted by the author, Düsseldorf, 21 June 2012, 27:21 min

customer will continue to use the service or not. Vodafone's approach is encapsulated in its customer relations' motto "Making customers into fans". Their goal is to give users sneak previews of the service, so that they can experience it personally and try it out, before they actually register.

Of course, when it comes to payment services, there are severe limits to how much one can use the service without committing to it, since it is impossible to avoid identifying the user when one makes a digital money transfer. Usability is closely linked to standards, since standardized technical interfaces prescribe specific procedures. There are certain learned ways of interacting with digital user interfaces that are well established and have already been highly optimized for user friendliness. These include registration procedures and automated data input procedures. With mobile terminals, in particular, it's advisable to incorporate learned native interaction features specific to the operating system and device concerned.

Fengler also describes the way in which viral effects can help to remove or bridge risks. When these are in operation, users prefer not to research the new technical solution in depth themselves. Instead, they prefer to observe their environment and evaluate the behaviour of others. The opinion of someone from his social circle and neutral observation of large numbers of people at the supermarket can both make an impact on the customer and influence his general attitude towards the new service. Sociologists call this phenomenon generalization. Generalization serves to simplify social processes and, just like trust, it reduces complexity.[222] If a service is broadly and generally accepted, this will have a positive influence on its acceptance by a new user. The establishment of a mobile payment service can be intricately connected with the social phenomenon of personal recommendations.

3.4 PayPal

PayPal was founded in 1988 and is now one of the leading suppliers of digital payment services worldwide.[223] It was a subsidiary of eBay and originally grew organically in tandem with the mother company.

3.4.1 The Company

PayPal's business model is now separate from that of its mother company and it has been delivering significantly higher growth rates than the online auction business. In the second quarter of 2012 alone, PayPal's year-on-year turnover grew by 26 percent to USD 1.36 billion.[224] PayPal now represents 38 percent of eBay's total annual turnover.[225]

---

[222] Cf. Luhmann, 2005, p.153
[223] Cf. PayPal (2012): Quarterly report Q2 2012. Fast Facts
[224] Cf. PayPal (2012): Quarterly report Q2 2012. Fast Facts
[225] Cf. Financial Times Deutschland: (18.04.12): Ebay wächst rasant dank Bezahltochter Paypal; PayPal (2012): Quarterly report Q2 2012. Fast Facts

As a digital payment service, PayPal permits international money transfers without the exchange of payment details between sender and recipient.[226] To do this, PayPal now administers 230 million active accounts in 190 countries.[227] In 21 of those countries, including Germany, PayPal provides a local website. PayPal's European branch only received its banking license in 2007, from the Luxembourgian regulatory agency for financial services (Commission de Surveillance du Secteur Financier, CSSF), allowing PayPal to handle payments in Europe.

PayPal entered the market with a mobile payment solution as early as 1998. At that time, the mobile solution transferred money from one Palm Pilot to another. In the following years, the company, which was ahead of its time, concentrated on the more profitable business of money transfers in online commerce and has now returned to its roots: to mobile payment.

In fall 2010, PayPal introduced mobile payment to the US.[228] It entered the German market a few months later, in March of 2011. Since then, worldwide mobile payment turnover has grown. From 2010 to 2011 alone, total mobile payment

---

[226] Cf. PayPal: So funktioniert PayPal,
[227] Cf. PayPal: About PayPal
[228] Cf. Spielberg, Holger (2012). Personal interview, conducted by the author, Dreilinden, 29 June 2012, 2:25 min

turnover has more than quadrupled and grew from USD 750 million to USD 4 billion.[229] Turnover is expected to almost double to reach USD 7 billion, in 2012.

3.4.2 Mobile Payment: A Center Piece

Holger Spielberg, Head of Mobile Payment & Innovation with PayPal Deutschland, states, "Mobile payment is at the heart of what we are doing here."[230] PayPal has already developed a number of mobile payment services. Their mobile solutions include Mobile Express Checkout, the PayPal App, PayPal Wallet, PayPal Here and QRShopping.

Mobile Checkout is part of PayPal's mobile-optimized online payment process. The customer can use it to stay on a merchant's mobile website and pay by PayPal from there. This payment process now accounts for over 90 percent of PayPal's total mobile payment turnover.[231]

PayPal has also developed a mobile app for use with iPhone, Android and Blackberry devices. The app allows users to administer their personal PayPal

---

[229] Cf. PayPal: Mobile Fast Facts
[230] Cf. Spielberg, Holger (2012). Personal interview, conducted by the author, Dreilinden, 29 June 2012, 0:21 min
[231] Cf. Spielberg, Holger (2012). Personal interview, conducted by the author, Dreilinden, 29 June 2012, 5:39 min

accounts and send and receive money while out and about.[232]

At the beginning of 2012, PayPal announced that they were developing a mobile wallet solution that was to be launched on the market that same year.[233] The aim is to develop new options for digital and/or mobile money transfers and flexible payment models -- such as payment in instalments, personal shopping lists with integrated price comparisons, local bargain search functions or settings for spending habits – and to integrate these features into the wallet solution.[234]

According to a survey by the Carlisle & Gallagher Consulting Group, final customers trust PayPal's mobile wallet solution more than those of Google or Apple because PayPal is a well-known digital payment provider.[235] They probably credit PayPal with more competence and integrity in this field than Google or Apple. Spielberg agrees, if you want to offer mobile payment, you must first know how to handle payment in general.'[236] He points to the experience handling payments which PayPal has

---

[232] Cf. iTunes Store: PayPal App, Google Play: PayPal App
[233] Cf. PayPal Blog (09.03.12): SXSW: PayPal to Give Attendees First Look at New Digital Wallet
[234] Cf. PayPal Blog (09.03.12): SXSW: PayPal to Give Attendees First Look at New Digital Wallet
[235] Cf. Tode (05.06.12): Consumers prefer PayPal mobile wallet over Google and Apple
[236] Cf. Spielberg, Holger (2012). Personal interview, conducted by the author, Dreilinden, 29 June 2012, 1:50 min

accumulated over the past ten years and stresses that expertise in the areas of payment, transactions, scaling, safety and risk management is essential if you want to offer mobile payment.

PayPal has also developed the POS payment solution "PayPal Here" as a competitor to Square.[237] This payment solution allows stationary merchants to accept credit cards using an iPhone or android phone in combination with a card reader dongle.[238] The payment procedure is almost the same as the normal procedure for credit card payments and will therefore probably be more readily accepted by customers than payment using NFC technology. It is possible that customers perceive that NFC technology, because of its novelty, as less secure than the established swipe mechanism used for credit card payments.

The QR Shopping app is a further POS mobile payment solution.[239] The solution, introduced in March 2012, allows merchants to label their products with individual QR codes. This opens up some interesting possibilities for commerce since the QR code can be used in a shop window but also, just as easily, in a newspaper or an advertising poster.[240] The customer can use the mobile

---

[237] Cf. Des Marais 16.03.12): PayPal Here Takes Aim at Square
[238] Cf. PayPal: PayPal Here
[239] Cf. Spielberg, Holger (2012). Personal interview, conducted by the author, Dreilinden, 29 June 2012, 6:48 min
[240] Cf. Bender (01.03.12): PayPal kommt per QR-Code an den PoS

application to scan the QR code, the app will recognize the merchant's product, and the customer can pay for the product and it will be automatically shipped to the customer's delivery address.[241] This means that advertising space can be developed into a new sales channel. Discussing the transformation of advertising spaces into potential sales channels, Spielberg ascertains that "Without payment, it's just advertising. With payment, it's commerce."[242] PayPal therefore offers merchants not simply a tool for payments but also a conversion tool for further sales opportunities.

However, the high proportion of turnover represented by the Mobile Checkout solution implies that incremental innovations can expect a greater and faster take-up rate and inspire more customer confidence, since they incorporate established and learned payment processes. Incremental innovations may have little novelty value, but the associated risks are easier to estimate and this allows trust to be established more quickly.

3.4.3 On Trusted Intermediary

The trust factor also played a central role in PayPal's introduction into Germany. PayPal's management originally decided not to launch PayPal on the German market because the local free direct debit

---

[241] Cf. PayPal: PayPal QRShopping
[242] Spielberg, Holger (2012). Personal interview, conducted by the author, Dreilinden, 29 June 2012, 8:22 min

scheme would make their own business model superfluous. In addition, PayPal had positioned itself as an alternative to credit card payment in the marketplace and in Germany at that time credit cards had a market penetration of only 8 percent.[243] However, against all business sense, they decided to become active in Germany anyway because it was well known that the German consumer mentality was very conservative and therefore there was a particular need for a trustworthy payment system for online commerce. Spielberg says, " ... everything we have done to win a share of the German market was based on trust."[244] With 10 million active users, PayPal in Germany is slightly smaller than the Postbank.[245]

PayPal positioned itself as the secure online payment method. It offered customer and merchant protection to protect both parties from fraud. When a customer files a claim under the buyer protection provision, PayPal assumes the role of arbitrator. On the basis of the two parties' descriptions of the situation, PayPal evaluates whether there has been a failure or partial failure to provide goods and services and what kind of compensation or refund would be appropriate.

---

[243] Cf. Spielberg, Holger (2012). Personal interview, conducted by the author, Dreilinden, 29 June 2012, 22:03 min
[244] Spielberg, Holger (2012). Personal interview, conducted by the author, Dreilinden, 29 June 2012, 24:47 min
[245] Cf. Spielberg, Holger (2012). Personal interview, conducted by the author, Dreilinden, 29 June 2012, 0:25 min

As well as customers, merchants also require protection against fraudulent behaviour on the part of customers. Customer descriptions of failure or partial failure to provide goods and services are not always accurate. In these cases, PayPal also acts as a trusted intermediary between merchant and customer. These mechanisms enable PayPal to ensure the merchant's trustworthiness towards the customer and vice versa, the aim being to facilitate the establishment of a relationship of mutual trust.

3.4.4 Mobile First

Spielberg remarks that it is critical that customers regard a payment system as a low-engagement product.[246] Users don't want to understand in detail exactly how it functions on a technical level. They just want to be able to rely on the payment working.

In order to develop an ideal low-engagement product, PayPal follows the "Mobile First" model. The design principle begins with the smallest customer interface. That is currently the mobile phone. The "Mobile-First" approach is based on simplicity, speed, user experience and scalability.[247] These guidelines should lead to the development of

---

[246] Cf. Spielberg, Holger (2012). Personal interview, conducted by the author, Dreilinden, 29 June 2012, 8:36 min
[247] Cf. Spielberg, Holger (2012). Personal interview, conducted by the author, Dreilinden, 29 June 2012, 36:04 min

simple, fast and user-friendly mobile payment solutions.

A smooth payment process needs to prevent any moments of confusion or distrust on the part of the user, since, as Spielberg emphasizes, users aren't conscious of trust factors, but they are all the more aware of moments of distrust.[248] Merchants, on the other hand, evaluate trustworthiness consciously and concretely according to the transparency and/or the information exchanged between PayPal and the merchant.

These differences in perception can arise from the difference in the intensity of the interactions that take place between PayPal and the merchant as opposed to the customer. The customer wants a low-engagement payment solution, a quick process, since his attention is focused primarily on the purchase. The customer estimates the risks of the transaction according to the price of the payment and chooses a payment option. The merchant, on the other hand, decides to use PayPal on the basis of a business evaluation. He or she therefore examines the system's security and transparency before integrating it into his/her own procedures. This phase of conscious analysis means that trust factors are much more obvious criteria for the merchant.

---

[248] Cf. Spielberg, Holger (2012). Personal interview, conducted by the author, Dreilinden, 29 June 2012, 41:11 min

A negative reputation can cause both final customers and merchants to experience distrust in the service. PayPal's PR scandals have clearly demonstrated this. For example, in December 2010, PayPal closed Wikileaks' donations account.[249] Wikileaks sympathizers demonstrated their solidarity by closing their own PayPal accounts.[250] Dirk Rossmann, owner of the Rossmann pharmacy chain, also temporarily barred PayPal as a digital payment method, when PayPal opened accounts in Germany for merchants who sold Cuban rum and cigars and therefore indirectly got involved in American politics within the German state.[251] The loss of confidence was so great that PayPal reacted quickly and took legal, business and organizational measures to prevent similar cross-border interventions in the future

3.4.5 Trust in Communication

Security and trust are important parts of communication with final customers for PayPal. Their advertising slogan for 2010 was "safererer".[252] Their ads humorously pointed out some everyday

---

[249] Cf. Heise online (04.12.10): PayPal sperrt Spendenkonto von Wikileaks
[250] Cf. Spielberg, Holger (2012). Personal interview, conducted by the author, Dreilinden, 29 June 2012, 1:01:01 min
[251] Cf. Spielberg, Holger (2012). Personal interview, conducted by the author, Dreilinden, 29 June 2012, 1:04:55 min
[252] Cf. Werben&Verkaufen (09.12.10): "Sichererer" ist der Werbeclaim des Jahres 2010

risks and particularly emphasized the safety of PayPal's money transfer system.

In addition to its importance in communications with final customers, the safety factor is a mainstay of the company's brand value.

In addition, the centrality of the trust factor is evident in the company's name, which represents PayPal as "a friend of payment". This implies friendship with the user and indirectly ascribes the characteristics of friendship to client-customer relations. Since every friendship requires a certain degree of mutual trust, the trust factor can be seen as a fundamental characteristic of PayPal. A study by Valtin & Fatke examines the subject of "trust in friendship". Valtin & Fatke summarize four fundamental aspects of friendship.[253]

Trust within a friendship means, trusting oneself in the sense of a personal openness, willingness to reveal personal things in the confidence of being accepted, safety from exploitation, injury or betrayal, reliability in the sense of keeping one's word and being trusted with things of value.

However, by doing this, PayPal emphasizes the characteristics of interpersonal trust rather than trust in a system. This has the unavoidable effect of contrasting their theoretical discourse with the

---

[253] See Valtin/Fatke, 1997, p. 75

natural limits that the intensity of this trust has in practice, where, for example, openness, empathy and selfless helpfulness are concerned.

4 Conclusion

This paper has investigated the role that the trust-factor plays in the design of a mobile payment service and in its communication with the public. The introduction examined the connections between the phenomenon of trust and monetary processes and went on to discuss the form, functions and determinants of trust in the context of new mobile payment services. The paper then continues with examining mobile payment in detail and ended with a practical comparison based on three interviews with representatives of Deutsche Telekom, Vodafone and PayPal. The comparison took the form of semi-open interviews with leading contact persons of subject-specific business units.

With regard to the connections between trust and money processes, this paper describes, that the very existence of money and its payment processes depends on the user's trust in the functionality of the monetary system. Trust is an intangible social mechanism, whose purpose is to reduce social complexity by minimizing foreseeable risks and uncertainties and positively anticipating future events. At the same time, trust is shaped by personal histories and moulded by familiarity – in the sense of the known and habitual. Learned symbolism, habits and reputation with their recognition values have a signal effect on trust and distrust.

At the same time, trust is a dynamic phenomenon. Trustworthiness -- and, with it, the threshold between trust and distrust – is subject to constant testing in a continual feedback loop. At the same time, every new positive experience strengthens and rebuilds trust. As this thesis explicitly examines mobile payment solutions, it focuses on systemic trust as opposed to interpersonal trust. Systemic trust refers to trust between a person and a system. Because of the lack of interpersonal reciprocity in communications between a person and a system, systemic trust takes longer to establish than interpersonal trust.

Any communication is primarily limited to the fulfilment of established rules and obligations. Communication, which follows the rules, therefore encourages trust whereas unplanned behaviour arouses distrust. Superficial trust in a system can be established quickly and is also more resistant to disappointments since negative experiences often never become public knowledge but fizzle out in individual experiences. Publicity is therefore more important when establishing trust in systems, since a system will appear trustworthier, if more people use it. Therefore trust in a system also implies a generalized form of trust.

Since trust in its abstract sense is intangible, we can only evaluate its role and meaning on the basis of its perceptible manifestations and aspects. In the literature, many attributes have been ascribed to trust, such as expertise, reliability, fairness, confidentiality and honesty, which have varying degrees of importance depending on the object of trust.

The detailed examination of mobile payment reveals not only the complexity and variety of its manifestations but also the clear need for trustworthy systems. Market research reveals significant market acceptance of mobile systems, but also demonstrates that mobile payment applications with only incremental degrees of innovation are trusted more and that there is a marked demand for security in technical systems (data processing and transfer).

The qualitative interviews show that trust is a relevant factor in the system's success and that the service providers recognize this. The security factor, in particular, is a well-known aspect of trust. There is a difference between systemic and perceived security. The paper also discussed aspects of trust like predictability (in the sense of individual control, standards adhered to, transparency and familiarity), viral effects (in the sense of the system's diffusion or market acceptance) and the reputation of the service provider and its partners.

These factors are particularly relevant for the user, although it seems likely that the user is not consciously aware of these factors, but is more sensitive to these conditions in which they are not fulfilled and to the threshold between trust and distrust. For the merchant, by contrast, security and transparency are important, but so is reliability (in the sense of orderly risk management). The merchant probably makes a conscious evaluation of trust factors, since the decision to integrate a mobile payment solution into his business is a question of economics.

The investigation of this subject of analysis has demonstrated that it is crucial to take trust into account if a mobile payment service is to be successful in the marketplace. A mobile payment service provider should therefore begin by determining what the relevant trust factors are for his target group and should then take appropriate concrete measures to address these in the design of the service, communications (including branding) and technical development. The investigation of the factors involved shows that there are not only overlaps between the different factors but also significant differences in the perception of trust and the willingness to trust which vary by target group, culture, social milieu and payment amount. The differences between the customer's and the merchant's perceptions of trust are of particular relevance when designing measures to be taken.

The design of the service needs to take learned payment sequences into account (trust factor: predictability). Relevant measures also include the incorporation of specific feedback and evaluation functions (truth factor: transparency). Where communication is concerned, it is helpful to include factors like transparency, honesty and integrity in the tone or history of a marketing campaign. The integration of personal recommendations also helps to promote the viral effect. Open, reliable and competent customer service relations also foster trust. If mistakes are made, open, speedy and honest communication can prevent customers from crossing the threshold into distrust. In the area of technical developments, the implementation of state-of-the-art methods of data handling and encoding, the use of intelligent algorithms to facilitate orderly risk management and the technical certification of payment processes can all increase trust in the security of the system.

By following these guidelines, a mobile payment service provider will be able to create a trusted platform and ideally position itself as a trusted intermediary between customer and merchant. It is precisely the translation of trust factors into service functions, which highlights the credibility of the mobile payment supplier. Allowing partners in commerce to evaluate each other (e.g. in the form of evaluations) could be a possible additional feature and/or a trust-building communication feature - before and after a purchase - between the two parties to the transaction.

The mobile medium offers special opportunities for the intensification of the trust relationship between user and service. Perhaps such an intensification of the relationship of trust between the user and the mobile payment service would be possible that the relationship could progress from the CBT phase, through the KBT to the IBT phase (see 2.1.5.1). In order to reach the KBT level, the user would need to have a completely satisfactory experience of using the service with relevant and fast information – in an emergency direct customer contact can even be established. With the help of additional individual services, such as the monitoring of personal spending or saving habits through the payment service, a user might even reach the IBT level of trust. In order to measure these developments, we would first have to define specific scales of measurement. Because only if we can measure trust factors on the basis of defined values, we will be able to amplify them.

Endnotes

List of Abbreviations

&	And
%	Percent
2D	Two-dimensional
A	User
AG	Aktiengesellschaft
API	Application Programming Interface
App	Application
ARPU	Average Revenue per User
BI	Behavioural intention
BIS	Bank for International Settlements
C2C	Consumer-to-Consumer
ca.	Circa
CBT	Calculus Based Trust
cf.	Compare
cm	Centimetre
CSS	Cascading Style Sheets
CSSF	Commission de Surveillance du Secteur Financier
e.g.	For example
E-Commerce	Electronic Commerce
EC	Electronic Commerce
Ed.	Editor
EDGE	Enhanced Data Rates for Global Evolution
Et al.	Et alia
EUR	Euro
EZB	Europäische Zentralbank
f.	And the following
ff.	Following pages
Fig.	Figure
FM	Frequency modulation
G	Potential profit
G&D	Giesecke&Devrient
GB	Gigabyte

Gbit/s	Gigabyte per second
GfK	Gesellschaft für Konsumforschung
GHz	Gigahertz
GmbH	Gemeinschaft mit beschränkter Haftung
GPRS	General Packet Radio Service
GPS	Global Positioning System
GSM	Global System for Mobile Communications
GSMA	Global System of Mobile Communication Association
HSCSD	High Speed Circuit Switched Data
HSPA+	High Speed Packet Access
HTC	High Tech Computer Corporation
HTML	Hypertext Markup Language
IBT	Identification Based Trust
IDC	International Data Corporation
iOS	iPhone Operating System
ISO	International Organization for Standardization
IT	Information Technology
ITU	International Telecommunication Union
IVR	Interactive Voice Response
Kbit/s	Kilobit per second
KBT	Knowledge Based Trust
KHz	Kilohertz
KWG	Kreditwesengesetz
L	Potential lost
LTE	Long-Term-Evolution
M-Commerce	Mobile Commerce
M-Payment	Mobile Payment
Mbit/s	Megabit per second
MC	Mobile Commerce
MHz	Megahertz
Mio.	Million

MMS	Multimedia Messaging Service
MNO	Mobile Network Operator
MPSP	Mobile Payment Service Provider
Mrd.	Billion
MSISDN Digital	Mobile Station Integrated Services Network
MVNO	Mobile virtual network operator
mWallet	mobile Wallet
n. y.	no year
n. p.	no page
NFC	Near Field Communication
p	Chance of winning
p.	Page
P2P	Person to Person, Peer to Peer
PC	Personal Computer
PDA	Personal Digital Assistant
PEOU	Perceived Ease of Use
PIN	Personal identification number
PoS	Point of sale
ppi	Pixel per inch
PU	Perceived Usefulness
QR	Quick Response
RFID	Radio Frequency Identification
RIM	Research In Motion
SDK	Software Developer Kits
SEPA	Single Euro Payments Area
SIM	Subscriber Identity Module
SMA	Stationary Mergent Automat
SMP	Stationary Mergent Person
SMS	Short Message Service
SSL	Secure Sockets Layer
St.	Saint
TAM	Technology Acceptance Model
TAN	Transaction number
TCP/IP	Transmission Control Protocol/Internet Protocol
UK	United Kingdom

UMTS	Universal Mobile Telecommunications System
URL	Uniform Resource Locator
US	United States
USA	United States of America
USSD	Unstructured Supplementary Service Data
VAS	Value Added Services
WAP	Wireless Application Protocol
Wi-Fi	Wireless Fidelity
www	World Wide Web
US	United States

List of Literature

Books and Articles

Ali-Yahiya, T. (2011): Understanding LTE and its Performance; Springer Science + Business Media; New York, Dordrecht, Heidelberg, London.

Allan, A. (2012): Learning iOS Programming; O'Reilly Media Inc.; 2nd edition; Sebastopol.

Beck, U. (1992): Risk Society. Towards a New Modernity; Sage Publications; 1st edition; London.

Bentele, G. (1998): Vertrauen / Glaubwürdigkeit; in: Jarren, Ottfired; Sarcinelli, Ulrich; Saxer, Ulrich (Hrsg.): Politische Kommunikation in der demokratischen Gesellschaft: ein Handbuch mit Lexikonteil; Westdeutscher Verlag; Opladen/Wiesbaden.

Berghaus, M. (2003): Luhmann leicht gemacht; UTB Verlag; 2nd edition; Köln et.al.

Bohn, U. (2007): Vertrauen in Organisationen: Welchen Einfluss haben Reorganisationsmaßnahmen auf Vertrauensprozesse? Eine Fallstudie; Dissertation an der Ludwig-Maximilians-Universität; München.

Bornschier, V. (2001): Generalisiertes Vertrauen und die frühe Verbreitung der Internetnutzung im Gesellschaftsvergleich; In: Kölner Zeitschrift für Soziologie und Sozialpsychologie 26; volume 53; issue 2; p. 373-400.

Bourdieu, P. (1992): Die verborgenen Mechanismen der Macht; VSA-Verlag; Hamburg.

Brieskorn, N. (2009): Sozialphilosophie: eine Philosophie des gesellschaftlichen Lebens; Kohlhammer Verlag; Stuttgart.

Bruhn, M.; Esch, F.-R.; Langner, Tobias (2009): Handbuch Kommunikation: Grundlagen – Innovative Ansätze – Praktische Umsetzung; Gabler Verlag; Wiesbaden.

Büssing, A.; Moranz, C. (2003): Faktoren initialen Vertrauens: Untersuchung zur Vertrauensbildung im Kontext virtueller Kooperationsanbahnung, in: Berichte aus dem Lehrstuhl für Psychologie; issue 73; TU München; München.
Burt, R. S.; Knez, M. (1996): Trust and Third-Party Gossip, in: Kramer, R. M., Tyler, T. R. (Ed.): Trust in Organizations: Frontiers of Theory and Research; Sage Publications; Thousand Oaks; p. 68 - 89.

Coleman, J. S. (1982): The Asymmetric Society; Stracuse University Press; 1st edition; New York.

Coleman, J. S. (1991): Grundlagen der Sozialtheorie. Handlungen und Handlungssysteme; Scientia Nova; Band 1; Oldenburg et. al.

Cooper, G. (2002): The Mutable Mobile: Social Theory in the Wireless World, in: Brown, B.; Green, N.; Harper, R.: Wireless World; Springer; London; p. 19 – 31.

Costello, D. (2002): Mobile Payments: Preparing for the mCommerce Revolution; Trintech; March 2002; White Paper; San Mateo.

Davis, F.; Bagozzi, R.; Warshaw, P. (1989): User Acceptance of Computer Technology. A Comparison of Two Theoretical Models; Management Science; INFORMS; volume 35; edition 8; p. 982 - 1003.

Deutsch, M. (1962): Cooperation and Trust – Some Theoretical Notes; in: Jones, M. (Ed.): Nebraska Symposium on Motivation; University of Nebraska Press; Oxford; Sp 275 - 320.

Diekmann, A.; Voss, Th. (2004): Die Theorie rationalen Handelns. Stand und Perspektive; in: Diekmann, A.; Voss, Th. (Ed.): Rational-Choice-Theorie in den Socialwissenschaften. Anwendungen und Probleme; Wissenschaftsverlag; Oldenbourg et.al.

Einwiller, S. (2003): Vertrauen durch Reputation im elektronischen Handel; DUV Gabler Edition Wissenschaft; Wiesbaden.

Esser, H. (2002): Soziologie. Spezielle Grundlagen. Volume 5: Institutionen. Campus Verlag; Frankfurt am Main.

Esser, H. (1996): Soziologie. Allgemeine Grundlagen; Campus Verlag; 2nd edition; Frankfurt am Main.

European Central Bank (2001): Blue Book: Payment and securities settlement systems in the European Union; 3rd edition; Frankfurt.

Evanoff, R. J. (2006): Integration in Intercultural Ethics; International Journal of Intercultural Relations; volume 30; edition 4; p. 421 - 437.

Fröhlich, G.; Rehbein, B. (2009): Bourdieu-Handbuch. Leben - Werk - Wirkung. J.B. Metzler, Stuttgart und Weimar.

Fukuyama, F. (1995): Trust. The Social Virtues And The Creation Of Prosperity; The Free Press Paperbacks; New York.

Gambetta, D. (2001): Kann man dem Vertrauen vertrauen?, in: Hartmann, M.; Offe, C. (Ed.): Vertrauen. Die Grundlage des sozialen Zusammenhalts; Campus Verlag; Frankfurt am Main; p. 204 - 237.

Gerck, E. (2002): Trust as Qualified Reliance on Information; part 1; The COOK Report on Internet; volume 10; edition 10; January 2002.

Giddens, A. (1995a): Konsequenzen der Moderne; 2nd edition; Suhrkamp Verlag; Frankfurt am Main.

Giddens, A. (1995b): Politics, Sociology and Social Theory. Encounters with Classical and Contemporary Social Thought; Stanford University Press; Stanford.

Hartmann, M. (2008): Usability Untersuchung eines Internetauftrittes nach DIN EN ISO 9241. Am Praxisbeispiel der Firma MAFI Transport-Systeme GmbH; Diplomica Verlag; Hamburg.

Helm, R.; Gehrer, M. (2006): Antezedenzen des Aufbaus von Vertrauen im extensiven Kaufentscheidungsprozess. Eine empirische Analyse der Wirkung des ersten Eindrucks, in: Bauer, H. H. (Ed.): Konsumentenvertrauen: Konzepte und Anwendungen für ein nachhaltiges Kundenbindungsmanagement; Verlag Franz Vahlen; München.
Henkel, J. (2002). Mobile Payment; In: Silberer, G.; Wohlfahrt, J.; Wilhelm, T. (Ed.): Mobile Commerce. Grundlagen, Geschäftsmodelle, Erfolgsfaktoren; Gabler Verlag; Wiesbaden; p. 327- 352.

Hubert, F.; Meyer, F.; Vogel, J.; Zimmermann, J. (2009): Co-Branding als Konzept zur Stärkung von Marken. Eine empirische Analyse im Konsumgütermarkt; in: Gierl, H.; Helm, R.; Huber, F.; Sattler, H. (Ed.): volume 42; Josef Eul Verlag; Lohmar – Köln.

Kirchgässner, G. (1991): Homo Oeconomicus. Das ökonomische Modell individuellen Verhaltens und seine Anwendung in den Wirtschafts- und Sozialwissenschaften; Mohr Siebeck; Tübingen.

Kumbruck, Ch. (2000): Digitale Signaturen und Vertrauen; Arbeit; volume 9; edition 4; p. 105 - 118.

Kunde, A. (2001): Mobile Payment - Entwicklungsperspektiven mobiler elektronischer Zahlungssysteme; GRIN Verlag, München.

Kuß, A.; Tomczak, T. (2004): Käuferverhalten; 3rd edition; Lucius & Lucius Verlagsgesellschaft; Stuttgart.

Krause, D. (2005): Luhmann-Lexikon; 4th edition; Lucius & Lucius Verlagsgesellschaft, Stuttgart.

Lammer, T. (2004): Mobile Payment Systems: Grundlagen – Praxisbeispiele – Erfolgsstrategien; Studien Verlag; Innsbruck.

Latour, B. (1988): The Prince For Machines As Well As For Machinations. In: Elliot, B. (Ed.): Technology and Social Process; Edinburgh University Press; Edinburgh; p. 20 – 43.

Laucken, U. (2005): Explikation der umgangssprachlichen Bedeutung des Begriffs Vertrauen und ihre lebenspraktische Verwendung als semantisches Ordnungspotential, in Dernbach, B.; Meyer, M. (Ed.): Vertrauen und Glaubwürdigkeit. Interdisziplinäre Perspektiven; VS Verlag für Sozialwissenschaften; Wiesbaden.

Lewicki, R. J.; Bunker, B. B. (1995a): Trust in relationships: a model of development and decline, in: Bunker, B. B.; Rubin, J. Z. (Ed.): Conflict, Cooperation and Justice; Jossey-Bass; San Francisco; p. 133 – 174.

Lewicki, R. J.; Bunker, B. B. (1995b): Development and Maintaining Trust in Work Relationships, in: Kramer, R. M.; Tyler, T. R. (Ed.): Trust in Organizations. Frontiers of Theory and Research; Sage Publications; Thousand Oaks.

Loose, A.; Sydow, J. (1994): Vertrauen und Ökonomie in Netzwerkbeziehungen. Strukturationstheoretische Betrachtung; In: Sydow, J.; Windeler, A. (Ed.): Management interorganisationaler Beziehungen. Vertrauen, Kontrolle und Infromationstechnologie; Westdeutscher Verlag; Opladen; p. 160 - 193.

Luhmann, N. (2005): Soziologische Aufklärung 1. Ausätze zur Theorie sozialer Systeme; 7th edition; VS Verlag für Sozialwissenschaften; Wiesbaden.

Luhmann, N. (2000): Vertrauen. Ein Mechanismus der Reduktion sozialer Komlexität; 4. Auflage; Lucius & Lucius Verlagsgesellschaft; Stuttgart.

Mathissen, M. (2009): Die Principal-Agent-Theorie: Positive und normative Aspekte für die Praxis; 1. Auflage; IGEL Verlag; Hamburg.

Mead, G.H. (1938): The Philosophy of the Act; University of Chicago Press; Chicago.

Meerman Scott, D. (2011): The New Rules of Marketing and PR. How to use Social Media, Mobile Applications, Blogs, New Releases and Viral Marketing to Reach Buyers Directly; 3rd edition; John Wiley & Sons; New Jersey.

Miebach, B. (2006): Soziologische Handlungstheorie: Eine Einführung; 3rd edition; Springer VS - Verlag für Sozialwissenschaften; Wiesbaden.

Misztal, B. (1996): Trust in Modern Societies; Blackwell Publishers; Malden.

Morgan, R. M.; Hunt S. D. (1994): The Commitment Trust Theory of Relationship Marketing; in: Journal of Marketing; edition 58; p. 20 - 38.

Neumann, D. (2006): Internet-Zahlungssysteme für Händler und Verbraucher im deutschen Rechtssystem; in: Lammer, T. (Ed): Handbuch. E-Money, E-Payment & M-Payment; Physica-Verlag; Heidelberg.

Neumann, M. M. (2007): Konsumentenvertrauen, Messung, Determinanten, Konsequenzen; DUV Gabler Edition Wissenschaft; Wiesbaden.

Petermann, F. (1996): Psychologie des Vertrauens; 3rd edition; Hogrefe Verlag; Göttingen.

Pfeifer, W. (1993): Etymologisches Wörterbuch des Deutschen; 2nd edition; Deutscher Taschenbuch Verlag; München.

Picot, A.; Reichwald, R.; Wigand, R. T. (2003): Die grenzenlose Unternehmung: Information, Organisation und Management: Lehrbuch zur Unternehmensführung im Informationszeitalter; 5th edition; Gabler Verlag; Wiesbaden.
Pousttchi, K. (2004): Mobile Payment in Deutschland. Szenarienübergreifendes Referenzmodell für mobile Bezahlvorgänge; Deutscher Universitäts Verlag; Wiesbaden.

Ropohl, G. (1999): Allgemeine Technologie. Eine Systemtheorie der Technik; 2nd edition; Fachbuchverlag Leipzig; Leipzig.

Rotter, J. B. (1980). Interpersonal Trust, Trustworthiness and Gullibility; American Psychologist; volume 35; 1st edition; p. 1 – 7.

Schäfer, R. (2008): Mobile E-Mail: Erkennung mobiler Nutzer und Darstellung von E-Mails werden immer wichtiger. – In: Schwarz, T. (Hrsg.): Online-Marketing Beratungsbrief; Schimmel Media Verlag; Würzburg.

Simmel, G. (1922): Soziologie: Untersuchungen über die Formen der Vergesellschaftung; 2nd edition; Duncker & Humblot; München – Leipzig.

Smith, S. L. (2007): Gone in a Blink: the Overlooked Privacy Problems Caused by Contactless Payment Systems; Marquette Intellectual Property Law Review; volume 11; 1st edition; p. 218-219.

Strack, R. (1999): Sicherer kartenbasierter Zahlungsverkehr im Internet – Erfahrungen und Perspektiven, in: Hermann, A. ; Sauter, M. (Ed): Management-Handbuch Electronic Commerce – Grundlagen, Strategien, Praxisbeispiele; Vahlen Verlag; München.

Stuhlhofer, F. (1983): Unser Wissen verdoppelt sich alle 100 Jahre. Grundlegung einer „Wissensmessung"; Berichte zur Wissenschaftsgeschichte; volume 6, edition 1-4; p. 169 -193.

Sydow, J. (1998): Understanding the Constitution of Interorganizational Trust, in: Lane, Ch.; Bachmann, R. (Ed.): Trust Within and Between Organizations. Conceptual Issues and Empirical Applications; Oxford University Press; Oxford; p. 31 - 63.

Sztompka, P. (1999): Trust. A Sociological Theory; Cambridge University Press; Cambridge.

Taga, K.; Karlsson, J. (2004): Global m-payment report 2004: Making m-payments a reality; Arthur D. Little; Wien.

Tidwell, J. (2010): Designing Interfaces. Patterns of Effective Interaction Design; 2nd edition; O'Reilly Media Inc.; Sebastopol.

Teichmann, R.; Nonnenmacher, M.; Henkel, J. (2001): E-Commerce und E-Payment – Rahmenbedingungen – Infrastruktur – Perspektiven; Gabler Verlag; Wiesbaden.

Valtin, R.; Fatke, R. (1997): Freundschaft und Liebe. Persönliche Beziehungen im Ost/West- und im Geschlechtervergleich; Auer Verlag GmbH; Donauwörth.

Schweer, M.; Thies, B. (2003): Vertrauen als Organisationsprinzip; Hans Huber Verlag; Bern.

Szameitat, A. (2001): Einführung und Überblick. Die Chipkarte der deutschen Kreditwirtschaft – Einsatzmöglichkeiten – Chancen – Perspektiven; 1st edition; Bank-Verlag; Köln.

Thalmann, A. T. (2005): Risiko Elektrosmog: Wie ist das Wissen in der Grauzone zu kommunizieren? Weinheim: Beltz Verlag.

Trumpfheller, J. (2005): Kundenbindung in der Versicherungswirtschaft – Eine theoretische und empirische Analyse unter besonderer Berücksichtigung des Versicherungsvertriebs über Verischerungsintermediäre; VVW Verlag Versicherungswirtschaft; Karlsruhe.
Volken, T. (2002): Elemente des Vertrauens. Internetdiffusion in den Transformationsländern als Paradigmatest; Peter Lang Verlag; Bern.

Warren, R. (2012): Creating iOS 5 Apps: Develop and Design. Peachpit Press, Berkeley.

Weber, M. (1988): Über einige Kategorien der verstehenden Soziologie, in: Weber, M.: Gesammelte Aufsätze zu Wissenschaftslehre; 7th edition; Mohr Siebeck Verlag; Tübingen; p. 427 – 474.

Wiecker, M. (2002): Breitbandige, kabellose Übertragungstechnologie; in: Gora, W.; Röttger-Gerigk, S. (Ed.): Handbuch Mobile-Commerce; Springer Verlag; Berlin.

Wolff, M.-K. (2002): Marktchancen E-Payment, in: Sauerburger, H.: HMD – Praxis der Wirtschaftsinformatik. Zahlungssysteme / E-Banking; Band 224; volume 39; dpunkt Verlag; Heidelberg.

Zucker, L. G. (1986): Production of Trust: Institutional Sources of Economic Structure. 184-1920; in: Staw, B. M./Cummings, L. L. (Ed.): Research in Organizational Behavior; CT:JAI Press; Greenwich; p. 53 - 111.

Online

Apple: Mac OS X Developer Library. Overview of In-App Purchase, URL
https://developer.apple.com/library/mac/#documentation/NetworkingInternet/Conceptual/StoreKitGuide/APIOvervie w/OverviewoftheStoreKitAPI.html#//apple_ref/doc/uid/TP40008267-CH100-SW1 call-off 20.04.2012

Bender, H. (01.03.12): PayPal kommt per QR-Code an den POS, Der Handel, URL
http://www.derhandel.de/news/technik/pages/M-Payment-PayPal-kommt-per-QR-Code-an-den-POS-8307.html call-off 10.08.12

BIS (2000): Clearing and settlement arrangements for retail payments in selected countries, CPSS-Retail Report 2000, Committee on payment and settlement systems, Bank for international settlements, Basel, URL
http://www.bis.org/publ/cpss40.pdf call-off 21.04.12

Boston Consulting Group (2012): Bank profitability threatened by squeeze on payments business. URL
http://www.prnewswire.com/news-releases/bank-

profitability-threatened-by-squeeze-on-payments-business-74248937.html call-off 10.04.12.

Bremmer, M. (03.11.11): Native App oder mobile Website?, Computerwoche Online, URL http://www.computerwoche.de/netzwerke/mobile-wireless/2497516/ call-off 19.04.12

Buhan, D.; Cheong, Y. C. / Tan, L.-C. (2002): Telecom Media Networks. Mobile Payments in M-Commerce, Cap Gemini Ernst & Young.
https://docs.google.com/viewer?a=v&q=cache:jDqf3R8U0AsJ:www.ebusinessforum.gr/engine/index.php?op%3Dmodload%26modname%3DDownloads%26action%3Ddownloadsviewfile%26ctn%3D942%26language%3Del+Telekom+Media+Network+Mobile+Payments+in+M-Commerce&hl=de&gl=de&pid=bl&srcid=ADGEEShunel3Hucr1K4Okbebq1pIddoos4kV_D86-5WjcfSEanawQ0Ujh4phle5v_Z2VYXX5QLS23e_4sZRhlIBbOF6WnW-CG3VsU4VTgFJh37CffHKR97WvrpeoNvxqfgEymD6kLrW_&sig=AHIEtbRtmuWZ_9_JnWIt9Kvj0QNGDHpTqg call-off 21.04.12

Bundesnetzagentur (2012): Teilnehmerentwicklung im Mobilfunk nach Netzen pro Quartal
http://www.bundesnetzagentur.de/cln_1931/DE/Sachgebiete/Telekommunikation/Marktbeobachtung/Mobilfunkteilnehmer/Mobilfunkteilnehmer_Basepage.html call-off 08.07.12

Bundesnetzagentur (18.08.00): UMTS-Versteigerungsverfahren abgeschlossen - Gesamtsumme von 99,3682 Mrd. DM erzielt, URL
http://www.bundesnetzagentur.de/SharedDocs/Pressemitteilungen/DE/2000/000818UMTS-Versteigerung.html?nn=106984 call-off 04.05.12

Capgemini Consulting (2009): Mobile Payment Studie. Blaze the Way for Mobile Payment in Germany. Representative consumer survey and analysis of business models, URL http://www.google.de/url?sa=t&rct=j&q=Mobile+Payment+Capgemini+2009&source=web&cd=1&ved=0CEMQFjAA&url=http%3A%2F%2Fwww.de.capgemini.com%2Finsights%2Fpublikationen%2Fblaze-the-way-for-mobile-payment%2F%3Fd%3D8FC78A3E-0996-AD35-7311-EB99EA109DCA&ei=zt2WT8fLFozvsgbMyfDfDQ&usg=AFQjCNGtj2ta0FwZMCwGdSMqsJ19lofYew&cad=rja call-off 27.04.12

CeBIT Leitthema (2012): „Managing Trust", Statements aus der Branche zu "Managing Trust". "Vertrauen und Sicherheit in der digitalen Welt", URL http://www.cebit.de/de/ueber-die-messe/themen-und-trends/news/managing-trust/statements-aus-der-branche-zu-managing-trust call-off 04.05.12

Chip Online (12.09.11): App-Downloads: Android überholt Apple noch 2011, URL http://business.chip.de/news/App-Downloads-Android-ueberholt-Apple-noch-2011_51603208.html call-off 05.05.12

Chip Online (28.12.09): So bekommen Sie Ihr Geld zurück. So werden Sie Klingelton-Abos los, URL http://www.chip.de/artikel/So-werden-Sie-Klingelton-Abos-los-4_32659805.html call-off 20.04.12

Colao (20.09.11): Vodafone Group Präsentation. Bringing „Supermobile" to life, URL http://www.vodafone.com/content/dam/vodafone/investors/conference_presentations/bernstein_200911.pdf call-off 08.08.12

Computer Bild.de (16.09.11): Neue Facebook-Einstellung: Freund, Fan oder Abonnent?, URL http://www.computerbild.de/artikel/cb-Aktuell-Internet-Neue-Facebook-Einstellung-Freund-Fan-oder-Abonnent-6447468.html call-off 05.07.12

Computerwoche (14.03.12): Europas Telekomriesen unter Verdacht, URL http://www.computerwoche.de/management/compliance-recht/2506993/, call-off 03.08.12

Contius, R.; Martigoni, R. (n.y.): Mobile Payment im Spannungsfeld von Ungewissheit und Notwendigkeit, URL http://subs.emis.de/LNI/Proceedings/Proceedings25/GI-Proceedings.25-5.pdf call-off 17.04.12

DesMarais, Ch. (16.03.12): PayPal Here Takes Aim at Square, URL http://www.inc.com/christina-desmarais/mobile-payment-platforms-paypal-vs-square.html call-off 10.08.12

Deutsche Bundesbank: Informationen über die Sicherheitsmerkmale der Euro-Banknoten, URL http://www.bundesbank.de/bargeld/bargeld_banknoten.php call-off 02.05.12

Deutsche Bundesbank: SEPA – Der einheitliche Eurozahlungsraum, URL http://www.bundesbank.de/zahlungsverkehr/zahlungsverkehr_sepa.php call-off 05.05.12

Deutsche Telekom (02.07.12): Mastercard und Deutsche Telekom, URL http://www.telekom.com/medien/konzern/132532 call-off 01.08.12

Deutsche Telekom (02.07.12): Studie Mobile Payment, URL http://www.telekom.com/medien/medienmappen/mobile-payment/132524 call-off 01.08.12

Deutsche Telekom: Geschäftsbericht 2011, URL www.telekom.com/static/-/102686/12/120223-gb11-pdf-si call-off 29.07.12

Deutsche Telekom: Mobiles Bezahlen, URL http://www.telekom.com/medien/medienmappen/mobile-payment/132524 call-off 01.08.12

Deutsche Telekom (02.07.12): Telekom stellt sich für den Payment-Markt auf, URL http://www.telekom.com/medien/medienmappen/mobile-payment/132530 call-off 01.08.12

Deutsche Telekom: Unternehmenspräsentation 2012, URL www.telekom.com/static/-/95406/6/unternehmens_praesentation-si call-off 29.07.12

Donovan, J. (27.02.12): Texas Instruments Announces New Partnerships For OMAP 5, But Wait … There's More, TechCrunch, URL http://techcrunch.com/2012/02/27/texas-instruments-announces-new-partnerships-for-omap5-but-wait-theres-more/ call-off 05.05.12

Elektronik Kompendium: HSPA, URL http://www.elektronik-kompendium.de/sites/kom/1301141.htm call-off 04.05.12

European Payments Council (07.02.12): Mobile Payments. 2nd Edition, White Paper, URL http://www.europeanpaymentscouncil.eu/knowledge_bank_download.cfm?file=EPC492-09%20White%20Paper%20Mobile%20Payments%20version%203.0.pdf call-off 17.04.12

Financial Times Deutschland (18.04.12): Ebay wächst rasant dank Bezahltochter Paypal, URL http://www.ftd.de/it-medien/medien-internet/:auktionsportal-ebay-waechst-rasant-dank-bezahltochter-paypal/70024536.html call-off 08.08.12

Free Patents Online (23.08.11): United States Patent Application US20110269423, URL http://www.freepatentsonline.com/20110269423.pdf call-off 05.05.12

Gartner (15.02.12): Gartner Says Worldwide Smartphone Sales Soared in Fourth Quarter of 2011 With 47 Percent Growth, URL http://www.gartner.com/it/page.jsp?id=1924314 call-off 04.05.12

GfK Studie (12.05.11): Mobile Payments: Trust and familiarity are crucial in driving adoption, URL http://www.gfk.com/group/press_information/press_releases/007888/index.en.html call-off 04.05.12

Go Smart: 2012 always-in-touch, Studie zur Smartphone-Nutzung 2012, Otto Gruppe, URL http://www.ottogroup.com/media/docs/de/studien/go_smart.pdf call-off 07.05.12

Google Developers: In-App Payments, URL https://developers.google.com/in-app-payments/docs/ call-off 20.04.12

Google Play: PayPal App, URL https://play.google.com/store/apps/details?id=com.paypal.android.p2pmobile&hl=de call-off 10.08.2012

Google Wallet, URL http://www.google.com/wallet/ call-off 20.04.12

Graziano, D. (31.01.12): Apple taking NFC payments mainstream with iPhone 5, BGR, URL http://www.bgr.com/2012/01/31/apple-to-make-nfc-payments-mainstream-with-next-gen-iphone/ call-off 05.05.12

Greif, B. (16.08.11): Telekom, O2 und Vodafone treiben Handybezahldienst Mpass voran, ZDNet, URL http://www.zdnet.de/news/41555704/telekom-o2-und-vodafone-treiben-handybezahldienst-mpass-voran.htm call-off 20.04.12

GSMA (Juli 2010): Mobile Money for the Unbanked. Mobile Money Definitions, URL http://www.latinia.com/static/if/new/informes/Mobile_Money_Definitions.pdf call-off 04.06.12

Heise online (04.12.10): PayPal sperrt Spendenkonto von Wikileaks, URL http://www.heise.de/newsticker/meldung/PayPal-sperrt-Spendenkonto-von-Wikileaks-1147516.html call-off 11.08.12

Hilbert, M.; López, P. (01.04.11): The World's Technological Capacity to Store, Communicate, and Compute Information, In: Science Magazine, URL http://www.sciencemag.org/content/332/6025/60 call-off 03.06.12

Hillebrandt, F. (23.05.11): Bestes Smartphone 2011? Das Samsung Galaxy S2 im Smartphone-Test, URL http://www.smartphone-testportal.de/samsung-galaxy-s2-test/ call-off 04.04.12

IDC (06.02.12): Smartphone Market Hits All-Time Quarterly High Due To Seasonal Strength and Wider Variety of Offerings, URL http://www.idc.com/getdoc.jsp?containerId=prUS23299912 call-off 05.05.12

IDC (30.06.09): Smartphone Growth Encouraging. Yet the Worldwide Mobile Phone Market Still Expected To Shrink in 2009, URL

http://www.reuters.com/article/2009/07/30/idUS159870+3
0-Jul-2009+BW20090730 call-off 04.05.12

In-Stat (21.04.11): Annual Mobile Payment Transactions to
Reach 45 Billion in 2015, URL
http://www.instat.com/newmk.asp?ID=3110&SourceID=00
000512000000000000 call-off 05.05.12

ISO Standard. Search for 9241, URL
http://www.iso.org/iso/home/search.htm?qt=9241&publish
ed=on&active_tab=standards&sort_by=rel call-off 10.08.12

ITU (Dezember): Mobile cellular subscriptions. Key 2000-
2010 country data, URL http://www.itu.int/ITU-
D/ict/statistics/ call-off 13.06.12

iTunes Store: PayPal App, URL
http://itunes.apple.com/de/app/paypal/id283646709?mt=8
call-off 10.08.2012

Kaasinen, E. (2005): User acceptance of mobile services, VTT
Publications, ESPOO.
http://www.vtt.fi/inf/pdf/publications/2005/P566.pdf; call-
off 04.05.12

Khodawandi, D. ; Pousttchi, K. ; Wiedemann, D. (2003):
Akzeptanz mobiler Bezahlverfahren in Deutschland, URL
http://mpra.ub.uni-muenchen.de/3606/1/Akzeptanz-
mobiler-Bezahlverfahren_06-13.pdf call-off 19.04.12

Lawson, S. (27.02.12): Facebook Pushes for HTML5
Standardization, Mobile Payments, PC World Online, URL
http://www.pcworld.com/article/250766/facebook_pushes_f
or_html5_standardization_mobile_payments.html call-off
19.04.12

Little, A. D. (2009): Global M-Payment Report Update - 2009.
M-payments surging ahead: distinct opportunities in developed
and emerging markets, URL

http://www.adlittle.ch/uploads/tx_extthoughtleadership/AD L_Global_M_Payment_Report_Update_2009_Executive_Su mmary_01.pdf call-off 09.04.12.

Mansfield, I. (27.02.12): Vodafone Teams Up with Visa for Global Mobile Payments Push, cellular-news, URL http://www.cellular-news.com/story/53236.php call-off 05.05.12

Mpass, URL http://www.mpass.de/ call-off 20.04.12

mobile money l!ve (September 2011): M-Pesa the largest source of revenue for Vodafone in Kenya, URL http://www.mobilemoneylive.org/articles/m-pesa-the-largest-source-of-revenue-for-vodafone-in-kenya/16864/ call-off 04.05.12

mobiThinking (Februar 2012): Global mobile statistics 2012: all quality mobile marketing research, mobile Web stats, subscribers, ad revenue, usage, trends…; URL http://mobithinking.com/mobile-marketing-tools/latest-mobile-stats#subscribers call-off 04.05.12

Müller, M. (03.09.09): nokia money: Bezahlen mit dem Prepaid-Handy statt einem Konto, URL http://www.teltarif.de/nokia-money-prepaid-handy-bezahlen/news/35587.html call-off 05.05.12

NFC Phones: Several hardware solutions for Secure Contactless Payments, URL http://www.nfc-phones.org/secure-payments-with-nfc-phones-4-hardware-solutions/ call-off 05.05.12

Nokia Studie (1999): Mobile VAS, URL http://www.telecomsportal.com/Assets_papers/Wireless/Nokia_mobile_vas.pdf call-off 04.05.12.

o.V. (2003): How Much Infofmation?; URL http://www2.sims.berkeley.edu/research/projects/how-much-info-2003/printable_report.pdf call-off 04.05.12

Olivarez-Giles, N. (10.03.11): Square answers VeriFone's accusations on security of mobile credit card reader, Los Angeles Times, URL http://latimesblogs.latimes.com/technology/2011/03/square-answers-verifones-accusations-on-security-of-mobile-credit-card-reader.html call-off 05.05.12

PayPal (2012): Quartalsbericht Q2 2012. Fast Facts, URL https://www.paypal-media.com/assets/pdf/fact_sheet/PayPal_Q2_2012_Fast_Facts.pdf call-off 10.08.12

PayPal (15.03.12): PayPal Unveils PayPal Here: the First Global Mobile Payment Solution for Small Businesses, URL: https://www.paypal-media.com/press-releases/paypal-unveils-paypal-here call-off 03.05.12

PayPal (27.04.11): PayPal to Open Mobile Payments Library to Developers, URL https://www.thepaypalblog.com/2010/04/paypal-to-open-mobile-payments-library-to-developers/ call-off 05.05.12

PayPal: About PayPal, URL https://www.paypal-media.com/about call-off 10.08.12

PayPal: So funktioniert PayPal, URL https://www.paypal-deutschland.de/sicher-bezahlen/mobile-payment.html call-off 10.08.12

PayPal: Texting with PayPal, URL https://personal.paypal.com/us/cgi-bin/?cmd=_render-content&content_ID=marketing_us/mobile_text call-off 22.04.12

PayPal: Prepaid, URL https://www.paypal-prepaid.com/ call-off 19.04.12

PayPal: PayPal Here, URL https://www.paypal.com/webapps/mpp/credit-card-reader call-off 10.08.12

PayPal: PayPal QRShopping, URL https://www.paypal-deutschland.de/privatkunden/qrshopping.html call-off 10.08.12

PayPal: Mobile Fast Facts, URL https://www.paypal-media.com/assets/pdf/fact_sheet/PayPal_mobile_fast_facts.pdf, call-off 10.08.2012

PayPal Blog (09.03.12): SXSW: PayPal to Give Attendees First Look at New Digital Wallet, URL https://www.thepaypalblog.com/2012/03/sxsw-paypal-to-give-attendees-first-look-at-new-digital-wallet/ call-off 10.08.12

Prensky, M. (2001): Digital Natives, Digital Immigrants, URL http://www.marcprensky.com/writing/prensky%20-%20digital%20natives,%20digital%20immigrants%20-%20part1.pdf call-off 13.07.12

Rao, L. (05.06.11): Mobile Payments To Triple To $670B By 2015; Digital Goods Will Represent 40% Of Transactions; TechChrunch; URL http://techcrunch.com/2011/07/05/mobile-payments-to-triple-to-670b-by-2015-digital-goods-will-represent-40-of-transactions/ call-off 02.05.12

Reißmann, O. (19.05.12): Unsichtbares Kleingeld verrät seinen Besitzer, Spiegel Online, URL, http://www.spiegel.de/netzwelt/netzpolitik/sparkassen-pilotprojekt-kontaktlose-geldkarte-verraet-ihren-besitzer-a-831711.html, call-off 25.05.12

Reppekus, D. (2000): Sicherheitskonzepte des UMTS Standards, URL http://www.fernuni-hagen.de/NT/kurse/sem_2000/umts.pdf call-off 04.05.12

Rupf-Schreiber, M. (2006): Identifikation und Vertrauen in Organisationen. Eine empirische Untersuchung in der Bankenbranche, Freiburg, URL http://ethesis.unifr.ch/theses/downloads.php?file=RupfSchreiberM.pdf call-off 07.05.12

Sawall, A. (25.05.11): Googles Handy-Bezahlsystem wird morgen vorgestellt, golem.de, URL http://www.golem.de/1105/83720.html call-off 05.05.12

Schenk, E. (23.02.12): G&D präsentiert NFC-fähige SIM-Karten, elektroniknet.de, URL http://www.elektroniknet.de/kommunikation/news/article/86093/0/GD_praesentiert_NFC-faehige_SIM-Karten/ call-off 05.05.12

SevenOne Media (03.12.03): Langzeitstudie Mediennutzung in Deutschlande, vgl. URL http://www.wiwi-treff.de/home/index.php?mainkatid=1&ukatid=1&sid=9&artikelid=1353&pagenr=0 call-off 08.05.12

Sommer, S. (21.12.11): Telefonkonzern kauft sich eigene Bank, Manager Magazin Online, URL http://www.manager-magazin.de/unternehmen/it/0,2828,804627,00.html call-off 10.04.12

Square: Security, URL https://squareup.com/security call-off 05.05.12

Statista: Jährliche Wachstumsrate für bargeldlose Transaktionen von 2001 bis 2007 nach den 10 größten bargeldlosen Märkten, URL: http://de.statista.com/statistik/daten/studie/71272/umfrage

/jaehrliche-wachstumsrate-bargeldloser-transaktionen-nach-regionen/ call-off 04.05.12

Telekom Innovation Laboratories: Mobile Wallet, URL http://www.laboratories.telekom.com/public/Deutsch/Innovation/CeBIT_2012/Pages/mWallet.aspx call-off 02.08.12

Telekom Innovation Laboratories: Mobile Wallet - Die digitalisierte Brieftasche, URL http://www.laboratories.telekom.com/public/Deutsch/Innovation/success-stories/Pages/Mobile-Wallet.aspx call-off 05.08.2012

The Netsize Guide 2011: Truly Mobile, Gemalto, URL http://www.netsize.com/Netsize-Guide-TrulyMobile.htm call-off 07.05.12

Thomas, D.; Kwong, R.; Song, J.; Taylor, P. (16.12.11): Apple and Google in Christmas Showdown, Financial Times Online, URL http://www.ft.com/cms/s/2/6ecd4a0a-280a-11e1-a4c4-00144feabdc0.html#axzz1qWOE2mE6 call-off 05.05.12

Thomas, A. (2005): Vertrauen im interkullturellen Kontext aus Sicht der Psychologie, in: Mayer, J. (Hrsg.): Die Rolle von Vertrauen in Unternehmensplanung und Regionalentwicklung; München; forost Arbeitspapier Nr. 27; URL http://www.forost.lmu.de/fo_library/forost_Arbeitspapier_27.pdf call-off 07.05.12

Tiefenthäler, R. (05.03.12): Apple: Über 25 Milliarden Downloads im App Store, NotebookCheck, URL http://www.notebookcheck.com/Apple-UEber-25-Milliarden-Downloads-im-App-Store.70814.0.html call-off 20.04.12

Tode, C. (05.06.12): Consumers prefer PayPal mobile wallet over Google and Apple, URL http://www.mobilecommercedaily.com/2012/06/05/consum

ers-prefer-paypal-over-google-apple-for-mobile-wallet call-off 10.08.12

Touch&Travel, URL http://www.touchandtravel.de/ call-off 18.04.12

Venmo, URL https://venmo.com/ call-off 20.04.12

Venmo: What is "Trust" and why would I use it?, URL https://help.venmo.com/customer/portal/articles/659156-what-is-%22trust%22-and-why-would-i-use-it- call-off 05.05.12

Visa Pressemitteilung (05.03.12): Vodafone und Visa gehen weltgrößte Partnerschaft beim mobilen Bezahlen ein, URL http://www.visa.de/de/uber_visa_old/presse/aktuelle_presse mitteilungen/vodafone_und_visa_gehen_weltgr.aspx call-off 06.08.12

Visa: Visa Mobile Payment Pilots. Mobile Fact Sheet Chart, URL: http://www.corporate.visa.com/_media/visa-mobile-fact-sheet-cartes.pdf call-off 02.05.12

Vodafone Deutschland: Mobil bezahlen, URL http://www.vodafone.de/privat/service/mobiles-bezahlen.html call-off 15.07.12

Vodafone Deutschland: mpass – die sichere Bezahlmethode per Handy, URL http://www.vodafone.de/privat/service/mpass.html call-off 15.07.12

Vodafone Group Plc. (31.03.12): Annual Report 2011, URL http://www.vodafone.com/content/dam/vodafone/investor s/annual_reports/annual_report_accounts_2011.pdf call-off 13.07.12

Vodafone Pressemitteilung (27.02.12): Vodafone and Visa Announce World's Largest Mobile Payments Partnership, URL http://www.vodafone.com/content/index/media/news/visa_partnership.html call-off 06.08.12

Werben&Verkaufen (09.12.10): "Sicherererer" ist der Werbeclaim des Jahres 2010, URL http://www.wuv.de/nachrichten/unternehmen/sicherererer_ist_der_werbeclaim_des_jahres_2010 call-off 11.08.12

Wiedemann, D; Goeke, L; Pousttchi, K. (o. J.): Ausgestalltung mobiler Bezahlverfahren – Ergebnisse der Studie MP3, Arbeitsgruppe Mobile Commerce, Lehrstuhl für Wirtschaftsinformatik und Systems Engerneering, Universität Augsburg, URL http://www.wi-mobile.de/fileadmin/Papers/MP/Ausgestaltung-mobiler-Bezahlverfahren_71-09.pdf call-off 05.05.12

www.ingramcontent.com/pod-product-compliance
Lightning Source LLC
Chambersburg PA
CBHW050058230526
45470CB00004B/1578